Clara

W9-AET-291

July 5, 1978

Also by Jack Stoneley and A. T. Lawton

Is ANYONE OUT THERE?
C.E.T.I.

Cauldron of Hell: TUNGUSKA

JACK STONELEY

Scientific Editor: A. T. Lawton, F.R.A.S

SIMON AND SCHUSTER
NEW YORK

Published by Simon and Schuster
A Division of Gulf & Western Corporation
Simon & Schuster Building
Rockefeller Center
1230 Avenue of the Americas
New York, New York 10020

Designed by Elizabeth Woll
Manufactured in the United States of America

1   2   3   4   5   6   7   8   9   10

Library of Congress Cataloging in Publication Data

Stoneley, Jack.
    Cauldron of hell.

    Bibliography: p.
    Includes index.
    1. Tunguska meteorite.   2. Life on other planets.   3. Space
probes.   I. Lawton, Anthony T.   II. Title.
QB756.T8S85   957′.5   77-15590

ISBN 0-671-22943-5

Contents

5

Introduction

It happened seventeen minutes and eleven seconds after midnight * on the last day of June 1908. Suddenly—like the sun falling from the sky—a monster invaded our planet. A monster that ripped into the atmosphere and exploded with the appalling fury of a thirty-megaton nuclear bomb that could be heard seven hundred and fifty miles away. The colossal shock wave that followed reverberated twice round the world and was clearly recorded in Britain.

And yet no one had the remotest idea what had happened. No one, that is, except a few terrified peasants in an isolated region of central Siberia, who crossed themselves and were convinced it was the end of the world.

For it was here, in the uninhabited upper reaches of the River Podkamennaya (Stony) Tunguska, that this cosmic vandal—still unidentified seventy years later— shattered itself into eternity and flattened nearly eight hundred square miles of forest like a gale ranging through a cornfield, wrenching our mighty tree trunks by their roots. Its scorching winds seared vast areas of earth and swept great sheets of water from the rivers. Herds of

* Greenwich Mean Time.

reindeer were consumed and scattered about as charred corpses. Even forty miles away men were thrown into the air, and one horror-stricken herdsman was left speechless for seven years. And from this same Godforsaken outpost, 2,200 miles from Moscow, millions of tons of mysterious silvery dust fanned half across the globe, suddenly turning night into day.

One particularly frightening aspect of the monster's assault is that had its course varied by only a few degrees it would have exploded over the very heart of bustling Edwardian London.

Yet it was to be thirteen years before the first news filtered through to the West that something "out of this world" had taken place in Tunguska in 1908. The Imperial Tsarist government of Nicholas II sneeringly dismissed Siberian newspaper reports as irresponsible hearsay. Peasants who talked to correspondents were written off as imbeciles.

Even now, after numerous expeditions by Russian researchers, the Tunguska Event is still one of the world's most intriguing unsolved mysteries. For, whatever it was that caused this incredible explosion, no evidence of the original form has ever been found. It came, it destroyed and it vanished. It left no fragments, nor is there any sign of a crater.

Modern scientific techniques are revealing other strange phenomena. One is now being investigated by a Russian expedition that went to Tunguska in the summer of 1976. Layers of moss, calculated to have grown in 1908, have been found to contain particles *the composition of which is unknown anywhere else on this Earth.* Even more remarkable are the bizarre biological changes taking place. The rate of genetic mutation of flowers and trees there has speeded up an astounding twelvefold. There is now a flourishing coniferous forest in the area.

What *is* the hidden secret of Tunguska? For years experts had assumed it was a giant meteorite that fell, like the one that blasted a three-quarters-of-a-mile crater in Arizona. But, if this were so, what happened to it? There is no mighty scar of impact at Tunguska. There are no meteorite fragments. And why were some trees right in the center of it all left almost untouched?

The purpose of this book is to probe the intriguing mystery from the moment the monster struck to the many conflicting theories that are still being argued today. Part One contains the dramatic eyewitness reports and graphic accounts of dedicated men who searched the forbidding Tunguska region in the nineteen-twenties and -thirties. Part Two discusses some of the terrifying forces of nature that scientists have put forward as possible answers—such as exploding comets, fearsome "black holes" and the uncanny "antimatter" of outer space.

The final section poses the most intriguing question of all. Could the source of this monstrous power have been artificial? Could it conceivably have been created by an exploding artifact from some other world beyond our own solar system? Could it, in fact, have been an alien counterpart of the NASA Viking probe that touched down on Mars in July and September 1976? And, if so, could unexplained signals picked up by Russian scientists since 1973 be originating from a replacement satellite still orbiting our Earth?

Somehow scientists *must* solve the curious enigma of the Tunguska event. For we can never be sure it will not occur again—perhaps next time over the crowded streets of London, New York or Moscow.

Part One

1

Day of the Fireball

It is June 30, 1908. Stifled passengers on the dusty Trans-Siberian Railway train are still trying to doze after a clammy, restless night on the hard carriage seats. Others stare in bleary-eyed silence through the thick black smoke that coughs rhythmically from the engine up front, occasionally curling inside through the door cracks.

The view through the streaky grime of the compartment windows does little to break the monotony of a grueling journey. It is raw, untamed land: mile after rugged mile of naked plains, mountain ranges and gaunt regiments of forest trees.

A red-eyed man loosening a coarse and sticky shirt collar wonders what in hell's name convinced him to come out here. But he knows there's no turning back now. Like many of the other gray-faced passengers, he is traveling on a one-way ticket. He's one of the Tsarist Empire's twentieth-century pioneers. For, like the earlier covered-wagon trails of the wild American West, the Trans-Siberian Railway is the four-thousand-mile Russian artery to the East that will pump lifeblood into the deadest place on Earth.

Already, only two years after its completion in 1906,

a million hardy settlers have joined this trek into a lost world which, for centuries, has been nature's dungeon for the Tsar's political prisoners to sweat out their sentences in the mines or to freeze in the mighty forests. For Siberia sprawls like a giant wasteland over four and a half million square miles, from Vladivostok to the Urals and from China to the Arctic Ocean. It is half as big again as the United States.

Before the inauguration of the Trans-Siberian Railway, hordes of the banished—assassins, murderers, bandits and those who had incurred the displeasure of some government official, wretched individuals from all classes of nineteenth-century Russian society—dragged their chains across the Urals on foot. Sometimes the gilded carriages of pampered notables of the aristocracy, personally deported by the Tsar, joined the pathetic procession, often with their wives and children, their servants and what few family treasures they had been able to salvage. Later, Bolshevik leaders like Lenin and Stalin are to have their bitter taste of exile here.

Many of the new pioneering families will not survive the long and merciless Siberian winters when the temperature can vary fifty degrees Fahrenheit on the same day; when the earth freezes like granite and the rivers become tentacles of solid ice for eight long, weary months of the year. When a man's breath solidifies into crystals of ice and birds drop frozen stiff out of the sky. Only the fittest will survive in a wilderness where there is one doctor for every eight thousand square miles. Some of the less robust immigrants will even perish on the journey to this harsh new world. But to most of them, forbidding as it is, Siberia presents a personal challenge. To those willing and able to come to accept its rugged terms of survival, there is land, mineral wealth—and a kind of freedom they have not known under the insatiable Tsarist exploitation of European Russia.

Now a few Trans-Siberian passengers are glancing at their watches. It's a little after 7:15 A.M.* They'll soon

---

* Local time.

be rumbling into dreary little Filimonovo Junction. At least they can get out there and stretch their legs.

The sun is up. There is no cloud. Nothing stirs out there. It's like trundling across the desolate surface of the moon.

Then they see it—streaking across the blue sky like a bolt out of hell: an immense, searing ball of fire that dims even the blinding rays of the sun. Seconds later a towering tongue of flame licks the horizon and a thunderous explosion shakes the astounded passengers to their feet.

As the heavy coaches tremble on the metal track, the engine driver is convinced some of them have been derailed. He jams on the brakes hard. The train grinds fiercely to a shuddering halt as the travelers lurch forward, shake themselves and then scamper out onto the track.

For a few moments they listen to dull thuds and distant rumblings. Then nothing more. The air is still again. The mean, uninviting territory stretches lifelessly about them. They shudder, wondering what kind of satanic land they're venturing into.

But if those frightened travelers on the Trans-Siberian Railway had been 350 miles to the north, at the lonely little trading station of Vanavara huddled on the banks of the stony Tunguska River, they would most certainly have taken the next train back to civilization.

Here, where the hunting tribes of the nomadic Evenki people come to barter their furs for provisions, the sky itself appears to rip apart and burst into flame. The earth shakes with appalling violence. A searing blast of radiation scorches the handful of inhabitants, hurls men into the air, wrenches doors from their stout log cabins and shatters heavy iron stoves. In the fields a tent is lifted high into the air together with the Evenki hunters sheltering inside it. All around them cattle stampede and startled horses gallop off in all directions, dragging their plows behind them.

And yet the place where the monster struck is still thirty-seven miles away—deep in the mighty uninhabited

Yenisei Forest in the basin of the Tunguska River. Here the devastation is unbelievable. Whatever it is that unleashes its awesome power tears up tree trunks by the roots, tosses them into heaps like spilled matchsticks, severs others as cleanly as a lawn mower plowing through grass. Its hot breath belches forth, scorching everything before it for hundreds of square miles. All wildlife in its path, including herds of reindeer, are wiped out in those few disastrous moments. Had the center of the catastrophe been heavily populated, the death toll would have been phenomenal.

And yet it is to be twenty years before Western civilization learns the incredible things that have taken place deep in the no man's land of central Siberia on this hot summer morning in 1908, and to wonder at some of the uncanny phenomena that are to make it one of history's most intriguing unsolved mysteries. For the more searchingly scientists seek the elusive truth of the Tunguska holocaust, the stranger it is to become.

It is 5:30 A.M., Greenwich Mean Time, on June 30, 1908—almost five and a quarter hours after the Tunguska disaster. Three thousand seven hundred miles away, most of Edwardian London is still sleeping. But at three of the city's meteorological stations—Westminster, South Kensington and Shepherds Bush—the sensitive microbarographs suddenly begin picking up strange aerial shock waves. The same thing is happening at Cambridge, at Leighton in Bedfordshire and at Petersfield in Hampshire.

The recordings last for just under two minutes. The waves have, in fact, encircled the Earth twice before being picked up on their second journey. But they arouse little interest. No one bothers to find out from where they have originated. The most urgent truly British concern on this sunny Tuesday morning is to accurately forecast the weather for King Edward's visit to Newmarket races. Then there are the lawn tennis championships at Wimbledon and the all-important social event of the year, the Henley Regatta. A mass meeting of women suffragettes is planned at Caxton Hall, with a public rally in Parliament Square. They'll want to know if it's going to be a nice day, too.

A record of the anomalous seismic wave is nonchalantly filed away and forgotten, later to be catalogued merely as No. 1536 on the list of earthquakes for the year 1908.

But that night something very odd happens. It is abnormally light. The candy-floss sky over London is tinted with a peculiar bright-red glow, and late West End revelers jogging home in their hansom cabs are amazed to find they can actually read the small print in their newspapers at midnight. Not that they find any reference to the incredible Tunguska incident—the actual cause of the phenomena. The only tidbit of news from Russia that catches their eyes that morning is *The Times*'s story, headlined "Fatal Duel in St. Petersburg," telling how "Count Nicholas Sumaroff was shot in an encounter with Count Manteuffel of the Horse Guards" in a squabble over a woman.

Many other places also experience the strange "white night" of June 30 to July 1. In Glasgow, between 11 P.M. and 3 A.M., it is so light that all but the brightest stars are invisible. As in London, people can read clearly in the streets. At 10 P.M. throughout Germany groups of curious observers watch an unfamiliar greenish twilight which dramatically changes to a glowing red rim along the horizon at midnight. And everywhere the clouds are shining like silver. (At Greenwich Observatory an astronomer finds he cannot take photographs of stars because the unusual luminosity of the sky fogs the plates.)

In Antwerp and other Belgian cities reports describe the ruddy glow of a huge fire rising above the horizon that makes it possible to see the seconds hand of a watch. And in some places there is no night at all. A student from Kazan University on vacation three hundred miles east of Moscow is astounded to find he can take excellent photographs in the streets at midnight. These are published in the magazine *Meteoritika*. From Saratov on the River

Volga passing ships are clearly visible nearly two miles away.

The following brief items appear in the columns of *The Times* of London on Friday, July 3. They receive the same prominence as the court report of a motorcyclist driving without lights "at a furious pace, from 15 to 25 miles an hour."

BRITISH ASTRONOMICAL ASSOCIATION— At the monthly meeting held on Wednesday evening at Sion College, Victoria embankment, Mr. G. J. Newbegin drew attention to the disturbed state of the solar atmosphere, showing a drawing and giving a description of a very large prominence that he had observed and measured in the morning of that day (July 1) and that showed unusual changes of form. Allusion was made by Mr. E. W. Maunder and Mr. H. P. Hollis (both of the Royal Observatory) to the long-lasting aurora of the previous evening.

CURIOUS SUN EFFECTS AT NIGHT.
TO THE EDITOR OF THE TIMES.

Sir,—Struck with the unusual brightness of the heavens, the band of golfers staying here strolled towards the links at 11 o'clock last evening in order that they might obtain an uninterrupted view of the phenomenon. Looking northwards across the sea they found that the sky had the appearance of a dying sunset of exquisite beauty. This not only lasted but actually grew both in extent and intensity till 2.30 this morning, when driving clouds from the east obliterated the gorgeous colouring. I myself was aroused from sleep at 1.15, and so strong was the light at this hour, that I could read a book by it in my chamber quite comfortably. At 1.45 the whole sky, N. and N.E., was a delicate salmon pink, and the birds began their matutinal song. No doubt others will have noticed this phenomenon, but as Brancaster holds an almost unique position in facing north to

the sea, we who are staying here had the best possible
view of it.

Yours faithfully,
HOLCOMBE INGLEBY,
Dormy House Club, Brancaster, July 1.

On July 4, a leader article in *The Times* goes into the
phenomena more deeply.

THE RECENT NOCTURNAL GLOWS

The remarkable ruddy glows which have been seen
on many nights lately have attracted much attention
and have been seen over an area as far as Berlin.
There is considerable difference of opinion as to their
nature. Some hold that they are auroral; their colour
is quite consistent with this view, and there is also
the fact that Professor Fowler of South Kensington
predicted auroral displays at this time from his observations, which showed great disturbances in the
sun's prominences. There was a slight, but plainly
marked disturbance of the magnets on Tuesday night,
and this materially strengthened the auroral theory,
as the two phenomena are very closely correlated.
However, this was shaken on the following night,
when the glow was quite as strong, but the magnets
were exceptionally quiet.

This convinced many who had before been in-
clined to the auroral theory, that the phenomenon
was an abnormal twilight glow . . . It is only necessary to suppose that some temporary condition of the
atmosphere made this twilight much brighter and
redder than usual.

The article then goes on to compare the phenomenon
with that witnessed after the enormous Krakatoa Volcano
eruption in the East Indies a quarter of a century before,
when abnormal sunset glows continued for months. This
furious holocaust blasted a column of fiery dust thirty-
three miles high and scattered rocks like mammoth hail-
stones for fifty miles. The volcano collapsed into a
six-hundred-foot sea crater, causing 36,000 people and

nearly three hundred towns to be destroyed in the tidal wave that followed. (See Chapter 10.)

We may recall the circumstances of the wonderful glows which were seen in this country in the autumn of 1883, and which were due to the dust scattered in the upper atmosphere by the terrific outburst at Krakatoa at the end of August [goes on *The Times*]. These glows had many points in common with the recent ones; the deep, lurid colour suggesting a distant conflagration (many were for some time doubtful whether Tuesday's glow was not due to this cause), and both glows were seen at a much longer interval after sunset than ordinary sunset glows, and the latter had already faded before the abnormal glow began. This indicated an extraordinary height for the dust causing the glow, and consequently the extreme fineness of the latter.

By charting the places and dates of the first visibility of the glows in 1883, it was found that the dust was carried westward by a previously unknown upper current at a speed of some 80 miles an hour. It did not reach the British Isles till its third circuit of the globe, each circuit having a wider range in latitude. We thus see that distance is no obstacle in vast cosmical phenomena of this kind, which are absolutely world-embracing.

No volcanic outburst of abnormal violence has been reported lately; there have, however, been some moderate outbursts in the Pacific during the spring, and it is possible that the dust may have reached us from these, or from some unreported eruption in some little-known region of the world.

By July 5, the news of the strange white nights reaches *The New York Times* from its man in London.

LIKE DAWN AT MIDNIGHT, LONDON SEES SKY BLUE AND CLOUDS TIPPED WITH PINK AT THAT HOUR

LONDON, July 4: Several nights throughout the week were marked by strange atmospheric effects which Dr. Norman Lockyer, of the South Kensington Solar Physics Laboratory, believes to be a display of the aurora borealis. Following sunsets of exceptional beauty and twilight effects remarkable even for England, the northern sky at midnight became light blue, as if the dawn were breaking, and the clouds were touched with pink in so marked a fashion that police headquarters was rung up by several people who believed a big fire was raging in the north of London.

For the following two months the Northern Hemisphere continued to have brighter nights than normal. However, it was not until 1930 that this phenomenon, together with the hitherto unaccountable seismic recordings, was directly linked to the mysterious Tunguska Event. It was then that the eminent English meteorologist C. J. P. Cave realized that the dates coincided. Dr. F. J. W. Whipple, then superintendent of Kew Observatory, London, also noted that the Tunguska district lay approximately in a direct line with Cambridge and Petersfield, where the seismic waves had been picked up. Later research concluded that as the giant fireball thundered toward the Earth several million tons of pulverized material had scattered in the atmosphere and acted as a huge reflector to the rays of the sun, out of sight below the northern horizon. But exactly what this material was remained a mystery.

Within days of the Tunguska explosion, eyewitness reports begin to appear in a number of Siberian newspapers, though nothing is yet known of the incredible havoc that resulted. The nearest inhabited place from where the interviews come is the village of Kenzma, still 130 miles from the object's deadly point of impact. It is from here that the newspaper *Krasnoyarets* receives the following comprehensive report, published on July 13:

> An extraordinary atmospheric phenomenon was noticed in this region. A noise as from a strong wind was heard, followed immediately by a fearful crash, accompanied by a subterranean shock which caused

buildings to tremble. One had the impression that some huge beam or heavy stone had struck the buildings. This was followed by two equally forceful blows and an extraordinary underground roar like the sound of a number of trains passing simultaneously over rails. Then followed a sound like artillery fire—between fifty and sixty bangs, becoming gradually fainter.

Eyewitnesses relate that, before the first bangs were heard, a heavenly body of a fiery appearance cut across the sky. Neither its size nor shape could be made out, owing to its speed and unexpectedness. However, many people in different villages distinctly saw that when the flying object touched the horizon a huge flame shot up that cut the sky in two.

Bangs were heard as soon as the tongue of flames disappeared. On the island opposite the village, horses began to whinny and cows to low and run wildly about. One had the impression that the earth was about to gape open and everything would be swallowed up in the abyss. Fearful bangs resounded from somewhere, shaking the Earth, and the invisibility of the source inspired a kind of superstitious terror. People were literally dumbfounded.

One of the first news stories, in the Irkutsk paper *Sibir,* is from the village of Nizhne Karelinsk, more than two hundred miles from the Tunguska Basin, and is dated July 2. The correspondent writes:

> In the northwest, quite high above the horizon, the peasants saw a body shining very brightly (too bright for the naked eye) with a bluish-white light. . . . The body was in the form of a cylinder.
>
> The sky was cloudless, except that low down on the horizon in the direction in which this glowing body was observed, a small dark cloud was noticed. . . . It was hot and dry and when the shining body approached the ground it seemed to be pulverized, and in its place a huge cloud of black smoke was formed and a loud crash, not like thunder, but

as if from the fall of large stones or from gunfire, was heard.

All the buildings shook and, at the same time, a forked tongue of flame broke through the cloud. All the inhabitants of the village ran out into the street in panic. The old women wept and everyone thought the end of the world was approaching.

Some of these early reports of the happening are rather haphazard and conflicting. No one seems to know exactly where the object has landed. Many newspapers quickly dismiss the whole thing as local gossip.

One skeptical editor sends a reporter 350 miles to the town of Kansk, where he has been told a meteorite has fallen. Needless to say, as the town is nearly four hundred miles from the disaster zone, the investigator doesn't get very much of a story. The following contemptuous item appears in the newspaper on July 4:

> . . . The noise was considerable, but no stone fell. All the details of the fall of a meteorite here should be ascribed to the overactive imagination of impressionable people. There is no doubt that a meteorite fell, probably some distance away, but its huge mass and so on are very doubtful.

Other local editors continue to carry brief references. On August 14, the newspaper *Sibirskaya Zhizn* tells of the frightening effects the explosion has had on men working in a gold mine 150 miles away.

> . . . Everyone felt a trembling of the ground accompanied by a loud roar as though from thunder, followed immediately by two fainter crashes and then by at least ten more still fainter ones. The mine buildings creaked and groaned, the gold-washing machines quivered, and people ran out in fear onto the street. The horses fell on their knees. . . .

By the autumn of 1908 reports in the press dry up altogether. The cruel Siberian winter isolates the tiny hamlets and villages from even their closest neighbors. And a forbidding blanket of snow and ice covers the handiwork of the Tunguska monster.

One has only to consider the vastness of Siberia to appreciate how even the most shattering event could escape the notice of the outside world. In 1891, explorer and author George Kennan had written:

> You could take the whole of the United States of America from Maine to California and from Lake Superior to the Gulf of Mexico, and set it down in the middle of Siberia without touching anywhere the boundaries of the latter territory. You could then take Alaska and all the states of Europe, with the single exception of Russia, and fit them into the remaining margin like the pieces of a dissected map . . . and you would still have more than 300,000 square miles of Siberian territory to spare.

A famous eighteenth-century anecdote charmingly demonstrates the remoteness of far-eastern Siberia even from European Russia in the days before the Trans-Siberian Railway. It is recorded that the Empress Elizabeth Petrovna invited six virgins from the distant peninsula of Kamchatka to visit her at St. Petersburg, nine thousand miles away. They were chaperoned by an officer of the Imperial Guard. Before they had even reached Irkutsk, eastern Siberia's trading center, they are said to have each given birth to a baby—all fathered by their overindulgent military escort, who was severely reprimanded and replaced by another. Despite this precaution, by the time they got to St. Petersburg they had all conceived further offspring.

Jonas Stadling, a Swedish explorer at the turn of the century, describes the Siberian forest thus: "Dark and almost impenetrable, it covers a marshy lowland extending more than 4,000 miles from east to west, and from 1,000 to 1,500 miles from south to north. In this endless monotony there is no change, no variety. You may travel hundreds and thousands of miles without seeing a human habitation, or any living thing other than wild animals or perhaps some stray Tungus [Evenki]."

For only the native Evenki knows how to survive in the taiga. In return for his respect, the forest can provide him with food from its limitless larder of wildlife, plants and berries. Its herbs cure his ailments. Its valuable fur-coated creatures provide the pelts that keep him alive in winter. The natural instinctiveness of the birds gives him ample warnings of predators. His barometer for sudden changes in climate is the peculiar behavior of small animal and insect life. Here the Evenki can live and hunt for years without reemerging.

But the basin of the Podkamennaya Tunguska, one of the most deserted parts of the Siberian taiga, is shunned by even the majority of the Evenkis as a region they believe to be condemned by their ancient gods. It is easy, therefore, to understand how the most spine-chilling of legends could be manifested in the simple minds of these rugged yet peaceful people. For theirs is a world of witch doctors; of lost lakes where grotesque monsters dwell; of strange godlike immortal beings said to inhabit the frozen peaks of mountains. It is a land of spirit-seekers with supernatural powers. Some tribes believe that the ten-foot ivory tusks of mammoths found in a remarkable state of preservation in some deep-frozen areas of Siberia were those of a gigantic species of rat that lived in holes deep below the ground and was the cause of earthquakes.

In the world outside, the distant babblings of a few superstitious peasants are quickly forgotten. To the hierarchy at St. Petersburg—capital of the Tsarist regime, three thousand miles away—whatever was seen and heard in Tunguska concerns them no more than if it had been in the remote Scottish Highlands or the wild Australian Outback. They have far more pressing problems to worry about—such as the distant rumblings of a political volcano that could shatter their vast empire far more alarmingly than any fireball from the sky. They have more to fear from the growing wrath of the proletariat than from the wrath of the gods.

The most tortuous and bitter years in Russia's history are to follow. From 1914 to 1920 the country is to be ravaged by war, first in the marathon struggle against Germany and then three brutal, agonizing years of civil turmoil. It all claims a sickening twelve and a half million lives—almost eight percent of the Russian people. Three million more of them are to perish through starvation and disease. Little wonder that in 1921 few people bother themselves about vague thirteen-year-old rumors from the boggy pine forests of Tunguska.

But one man does. His name is Leonid A. Kulik.

2

Man with a Mission

A bearded man in his late thirties hugs the astrakhan collar of a greatcoat over his ears and listens to the clatter of heavy train wheels over the rails of another of the desolate little junctions strung along the Trans-Siberian Railway. It reminds him of machine-gun fire, so familiar to this young revolutionary during the long, savage years of war. But it is now mid-September 1921. The killing is over. There is time for men like Leonid Kulik to think of other things. And in the searching mind of this restless scientist one objective dominates: the Tunguska monster.

To Leonid Kulik from the Mineralogical Museum of St. Petersburg (now renamed Petrograd) the mysterious fireball of 1908 has become an obsession. Whatever it was, he is determined to seek it out.

As he has done countless times, he thumbs through a file of fraying newspaper clippings—those first vague reports of the strange affair. He wonders if the word of a handful of uneducated peasants so long ago really justifies his making this punishing journey into central Siberia.

Actually the original motive for the visit here had nothing to do with the Tunguska phenomenon. He hadn't even heard of it when he was commissioned by the Russian Academy of Sciences to collect general information about meteorite falls. But before his party of researchers left Petrograd a diligent Soviet researcher called Svyatsky had handed him a sheet ripped from a 1910 calendar. Pasted to the back of it was part of an article from the newspaper *Sibirskaya* printed in 1908. It described that Trans-Siberian journey on the morning of June 30, thirteen years ago, when the passengers at Filimonovo Junction had watched the fireball streak by. According to this report, later found to be incorrect, the driver and the passengers had found a red-hot meteorite that had cooled into a huge white stone block almost entirely buried in the ground.

This account had inspired Kulik to search the back numbers of other Siberian newspapers. The fearful stories, conflicting as many of them were, intrigued him. He had to find out what really happened.

Unfortunately this particularly ferocious winter of 1921 closes in early. By the time Leonid Kulik has calculated where the object struck, its violent secret is guarded by a vast fortress of ice and snow. But he doesn't waste those first precious months in Siberia. Instead he probes the still-vivid memories of many people in the towns, cities and villages in the populated areas of Kansk and Tomsk with persistent interviews and by distributing 2,500 searching questionnaires.

But, back in Petrograd, these reports—like those original newspaper accounts of 1908—are soon forgotten by the scientific fraternity. It is to be yet another six years before Kulik can pursuade the Russian Academy of Sciences to send his team out there again.

What finally does convince them is a succession of startling reports from other researchers working on various projects in the Tunguska region. They pass on stories told to them by the wandering Evenki people who were camped within twenty-five miles of the explosion. Stories of a thousand reindeer being wiped out, great forests being utterly destroyed and "water gushing up from the earth." One report comes from a geologist named Sobolev, employed by the Krasnoyarsk Museum near Kansk, who was working in the Tunguska area in the summer of 1923. There he met a man called Ilya Potapovich.

> Fifteen years ago [reads the geologist's account] Ilya Potapovich's brother lived on the River Chambé. One day a terrible explosion occurred, the force of which was so great that the forest was flattened for many versts * along both banks of the river.
> His brother's hut was flattened to the ground, its roof was carried away by the wind and most of his reindeer fled in fright. The noise deafened his brother and the shock caused him to suffer a long illness. In the flattened forest at one spot a pit was formed from which a stream flowed into the River Chambé. The Tunguska road had previously crossed this place, but it was now abandoned because it was blocked, impassable, and moreover the place aroused terror among the Tungusi people. From the Podkamennaya Tunguska River to this place and back was a three-day journey by reindeer.

In 1924 another geologist, S. V. Obruchev, also interviews Ilya Potapovich, who tells how his brother was "deprived of speech for several years" after the experience. Ilya himself was to be one of the guides for later expeditions by Kulik.

More fascinating eyewitness accounts are passed on to the Academy of Sciences by a member of the Russian Geographical Society, ethnographer I. M. Suslov, who, in March 1926, was working at the windswept little trading stations of Vanavara and Teterya on the Tunguska River. He too has closely interrogated the Evenki people. One of them is a woman called Akulina, the widow of Ilya Potapovich's brother. She describes the night of the great fireball like this: "Early in the morning, when everyone was asleep in the tent, it was blown up into the air, together with the occupants. When they fell back to earth the whole family suffered bruises." She and her husband lost con-

---

* A verst equals two thirds of a mile.

sciousness. When they came round they "heard a great deal of noise and saw the forest blazing around them and much of it devastated." An old man who was living with them at the time adds, "The ground shook and incredibly long pronounced roaring was heard. Everything round about was shrouded in smoke and fog from the burning falling trees."

Suslov also talked to sixty other Evenki people at the trading stations in June 1926. All agreed that when the fireball fell it "burned the trees, killed the dogs and the reindeer, injured people and flattened the taiga."

An engineer called Gundobin sends Kulik this story from Ivan Vasilyevich Kokorin, helmsman of a boat navigating the turbulent River Angora at the time: "In the north flashed up a bluish light and there passed in the sky (from the south) a fiery ball, considerably larger than the sun, which left behind it a wide luminous streak. Thereupon there occurred such a cannonade that all the laborers that were in the boat rushed to hide themselves in the cabin, having forgotten about the danger which threatened from the rapids. The strength of the sounds was so great that the oarsmen were entirely demoralized and it took great efforts to return them to their places in the boat."

Early in 1927 new research is also published in the magazine *Mirovedeniye* by the former director of the Irkutsk Magnetic and Meteorological Observatory, A. V. Voznesensky, who estimates that the thunderous explosion from Tunguska had been heard over 380,000 square miles. Like Kulik, he believes that the intruder was a meteorite. But, as there were a number of explosions, he concludes that it must have shattered into smaller bodies before landing. It is highly probable, he claims, that future investigators will find a mass of fragments.

Backed by this later information, together with warnings that the object might be destroyed by atmospheric corrosion, Kulik is eventually able to convince the academy that a further expedition to find the Tunguska monster is now justified. And so, in February 1927, Leonid Kulik finds himself back on the Trans-Siberian express to continue the incredible mission that is to dominate the rest of his life.

It is early spring in central Siberia, but still agonizingly cold as Kulik and his assistant Gyulikh load horse-drawn sledges with supplies of food and equipment at Taishet station, some one hundred miles east of Kansk.

Their next objective is the Vanavara trading station, 350 miles to the north. They travel through the village of Kezhma, where they replenish their supplies, and struggle on for three exhausting days into wild, uncharted country, finally staggering into Vanavara on March 25. It's like a miniature Dodge City before the prospectors came. Just a few log cabins and one or two primitive wooden buildings line the "streets" of ankle-deep mud.

At the trading station Kulik finds Ilya Potapovich, who agrees to guide them on horseback into the mighty taiga the following day. But the weary animals just can't make headway through the huge snowdrifts, and they have to turn back to Vanavara.

For the next few days Kulik gets to know some of the rugged residents of the outpost, as well as the tough Evenki hunters who come to the settlement to trade their furs. They discuss with dread the day they saw the sky "open to the ground and fire pour out—fire brighter than the sun," and how the ground trembled and earth thundered down onto their roofs. One man called Kosolapov describes how "a fierce heat scorched my ears." Another, Semenov, claims "a hot wind as if from a cannon" seemed to burn the shirt from his back as he sat in the porch of his house at the trading station.

"I saw a huge fireball that covered an enormous part of the sky," Semenov tells Kulik. "Afterward it became dark and at the same time I felt an explosion that threw me several feet from the porch. I lost consciousness for a few moments and when I came to I heard a noise that shook the whole house and nearly moved it off its foundations. The glass and the framing in the house shattered, and, in the middle of the area where the hut stands, a strip of ground split apart."

It all makes Kulik even more eager to penetrate the gaunt forest to find the thing that still strikes terror into so many Tunguskan hearts. He urges Ilya Potapovich to help them try again. The dubious man reluctantly agrees.

The staunch little party sets off, stumbling blindly through soggy, infection-ridden marshland, until they reach the last tiny outpost before what Kulik is told is the fringe of the stricken region. It is the cabin of a Tungusi called Okhchen, perched on the side of the Chambé River. Here they are given a night's shelter and reindeer to carry their belongings the rest of the way.

Two days later, on April 13, they cross the murky Makirta River that sweeps into the Chambé and, suddenly, the crippling hardships of the journey are forgotten. A weary but elated Leonid Kulik stands in awed silence on the river bank and gazes out across a nightmare of utter devastation. It is something quite beyond his wildest imaginings. Something sinister and unreal. He shudders in the thin icy air.

There, scattered like the huge gnarled bodies of a fallen army of giants, regiments of burly pines that had withstood centuries of violence from the most brutal climate in the world now lie flat and lifeless on the frozen ground. As he stares about him, Kulik can only wonder what kind of monstrous vandal could conceivably have caused such outrageous havoc, what unbelievable force must have raged across this forest nineteen years ago, stripping its majestic trees as naked as telegraph poles. What fury had torn these deep unyielding roots from the grasping earth and laid them bare?

The party push on for several days and reach the top of the high Khladni ridge, from which they can see for many miles across the ravaged landscape. From here it is even more breathtaking. Kulik scribbles his first dramatic impressions in his diary:

I cannot really take in the whole majestic picture.
A very hilly, almost mountainous region stretches
away tens of versts toward the northern horizon. In
the north, the distant hills along the River Khushmo
are covered with a white shroud of snow half a me-
ter thick. From our observation point no sign of
forest can be seen, for everything has been devastated
and burned, and around the edge of this dead area
the young, twenty-year-old forest growth has moved
forward furiously, seeking sunshine and life. One
has an uncanny feeling when one sees 20- to 30-inch-
thick giant trees snapped across like twigs, and their
tops hurled many meters away.

Kulik cannot wait to go on into this astounding graveyard littered with its corpses, the stumps of their amputated limbs groping out of the snow. For there, he is
sure, he will find the monster. But an unexpected problem
crops up. Ilya Potapovich and Okhchen have had enough.
They adamantly refuse to go any farther. They are convinced something supernatural still lurks in the shadows
of the great dead taiga. Something whose wrath they have
no wish to incite.

For Kulik it is a bitter blow. He has traveled more
than three thousand miles. He is now so close to the very
heart of this whole fantastic mission into the unknown.
But he realizes it would be suicidal to risk continuing
through the region with only his assistant Gyulikh. The
treacherous spring floods are due soon, and they are a
long way from the safety of the trading station. He has
no choice, therefore, but to return to Vanavara.

Fortunately, back here he meets some bold Evenki
hunters who are willing to take over as guides for a second
attempt, which begins on April 30. The snow is still
thick on the ground, but three days later they reach the
River Chambé and make camp. The ice is now beginning
to thaw and break up into thick chunky slabs that swirl
in the perilous rapids of the taiga rivers. Despite this,
Kulik decides to penetrate the devastated forest area by

sailing rafts along the Chambé and into the River Khushmo.

On May 19 the rafts are ready and the slow, tortuous cruise begins. Time and again they find the route blocked by massive ice floes. The water is rising alarmingly and surging over the river banks. A raft loaded with equipment rips from its overnight mooring and is waltzed downstream as easily as a fallen leaf in a rippling English stream. But, four days and twenty-five miles later, the party reach the Khushmo, and soon they are well into the devastation area.

Now they have to hack a way through great masses of uprooted trees blocking the river and sprawling along the banks.

It is May 30. The weary expedition has reached the mouth of the ill-tempered little River Churgima. Here, in the deep valley, they pitch their tents. Kulik senses he is within striking distance of the very center of the impact zone. He makes daily reconnaissances and finally decides that his objective lies beyond a barrier of ridges to the north. He pushes eagerly on.

Suddenly, spread out before him is a gigantic amphitheater of hills surrounding a huge marshy basin. The Evenkis know it as the legend-ridden Southern Swamp. To Kulik it resembles a mammoth cauldron, and this is what he will call it from now on: the Great Cauldron of Tunguska. It is the most stimulating experience of this dedicated young scientist's life. For somewhere inside it he is convinced he will find the burial ground of the cosmic brute that went berserk in this same month of June nineteen years ago.

Leonid Kulik's jubilation is expressed in his own notes after making a preliminary survey:

> There can be no doubt. I had circled the center of the fall! With a fiery stream of hot gases and cold solid bodies, the meteorite had struck the Cauldron with its hills, tundra and swamp, and, as a stream of water striking a flat surface splashes spray in all

directions, the stream of hot gases with the swarm of bodies penetrated the earth and, both directly and with explosive recoil, wrought all this mighty havoc.

He goes on:

> In the Cauldron there are hills, ridges, isolated summits and flat tundras, marsh, lakes and streams. The taiga, both in the Cauldron and outside it, has been practically destroyed by being completely flattened. All the former vegetation of the Cauldron, of the surrounding mountains and of a zone of several kilometers around them bears characteristic traces of uniformly continuous scorching, unlike the traces of an ordinary conflagration. The area with scorch marks is estimated to be tens of kilometers across.

As Kulik encircles the Cauldron, he is fascinated to find that the decapitated and uprooted trees have been scattered outward from the center, like the tangled spokes of a wheel—their exposed roots pointing to the hub, the tops facing outward in the direction they were flung in the incredible blast. Another remarkable discovery is that trees on the lower ground, which he is certain was the epicenter of the blast, are still standing. The farther he goes up the hillsides, the more severe the damage (see Figure 1). It's as though a giant scythe has been swept in a circle from the center. On the lower slopes, only the tips of the trees are missing. Higher up, the trunks are severed off—the remaining stumps gradually getting shorter and shorter until, at the summits, the trees are completely uprooted. Later studies are to show that only an explosion *above the ground* could have caused this pattern of devastation.

Kulik is still convinced the object that caused it all was a meteorite, especially when he finds numbers of flat holes of different sizes pockmarking the peat bogs like small lunar craters. He assumes that the body split up before landing, each fragment forming its own scar. After much later research he has to concede that these are naturally formed depressions in the earth caused by the thawing of the permafrost. No actual meteorite crater is found, nor is one ever to be found. But whatever it was that pounded from the sky had thrust the earth into strangely deformed folds, like a carpet that rumples up on a polished floor.

Kulik is overwhelmed by it all. There is so much more research he'd like to do. But, with enough food left for less than a week, he reluctantly leads his intrepid little group back toward Vanavara. He staggers up to the trading station nine days later, exhausted, ravenously hungry, but glowing with satisfaction. Leonid Kulik has found what he came for. He has walked in the Cauldron of Hell. Soon he must come back to find the thing that created it. He is not to know that the secret will still be hidden long after his death. For what he has done in the summer of 1927 is to present to the scientific world the riddle of the century.

3

The World Takes Notice

The mysterious Tunguska monster suddenly hits the headlines. Newspaper and magazine editors throughout the world begin poring over large-scale maps of central Siberia, prodding their fingers at faraway places they've never heard of until now. They rummage through their files. They call in their science writers. Why has it taken nearly twenty years to get onto this remarkable story?

The postrevolutionary hierarchy of Russia, eager to impress the West with a new progressive image, are now more than willing to listen to Leonid Kulik, who wastes no time in urging the Academy of Sciences to sponsor a further expedition. In December 1927 he submits his estimates, together with a letter to the presidium of the academy.

"The results of even a cursory examination," he tells them ecstatically, "exceeded all the tales of eyewitnesses and my wildest expectations. The devastation can only have been caused by an air wave of tremendous power." He then goes on firmly:

> For seven years I have been holding the view that since this fall occurred on the territory of the Soviet

Union, we are in duty bound to study it. If the matter was delayed until last year on the pretext that it was all pure fancy, this objection has now been swept away since the positive results of my expedition are irrefutable. Their unique scientific significance, like the significance of the Tunguska fall itself, will be fully appreciated only in history, and it is necessary to record all the remaining traces of this fall for posterity.

There are few objections to a new expedition. The scientific fraternity, skeptical for so many years, is now eager for more news. And so, on April 12, 1928, Kulik is back in Siberia.

The notorious spring floods are at their height when he arrives with zoologist V. Sytin at Vanavara trading station thirteen days later. Here they are joined by a movie cameraman from Moscow named Strukov who is to give the outside world its first dramatic glimpse of the rape of Tunguska. He almost records Kulik's death too.

When the party reaches the mouth of the River Chambé they have to travel upstream in boats towed by a pair of horses. Here the rapids are particularly treacherous, swirling and pounding the huge boulders that jut from the narrow riverbed. Suddenly Kulik's boat overturns. He is flung headlong into the icy water. The powerful current throws him against the rocks. Then, miraculously, as he is about to be swept helplessly downstream, his leg gets entangled in a mooring rope that has wrapped itself about a rock. He is able to struggle, gasping, to the surface and frantically grabs the side of the boat. Eventually he drags himself to the bank as the news-minded photographer calmly shoots the whole incident. Kulik, despite the ordeal, is still wearing his spectacles.

After building a temporary camp on the River Khushmo, near the mouth of the Churgima stream, with a shed perched on stilts to be out of reach of marauding wolves and bears, they finally settle in the heart of the Cauldron itself on June 23.

Kulik's main concern is to probe the small round holes

dotted about the area. He is still convinced they have been caused by meteorite fragments, and he's determined to find them. He digs deeply into the earth, but finds nothing. In fact, little new is to emerge from this 1928 visit. Its purpose is mainly to prepare the way for a later and far better equipped expedition. However, cameraman Strukov gets the first exclusive pictures that are to astound the world and help Kulik to get all the backing he needs for more extensive research.

Meanwhile, in the United States, Harvey H. Nininger of the Colorado Museum of Natural History, one of the country's foremost authorities on meteorites, tries to get the leading scientific societies there to finance a super-equipped American expedition to Tunguska to secure, as he puts it, "what is yet available of this greatest message from the depths of space that has ever reached our planet." All his efforts fail. In his book *Our Stone-Pelted Planet* he remarks:

> It is a sad comment on the mental alertness of the scientific world that even until now there has been no adequate effort put forth to collect the great fund of information which awaits any well-equipped expedition into these parts.
>
> Meanwhile, these same societies spend enormous sums studying volcanoes and other phenomena which are always with us, and worthy as are these studies, they could be pursued any time and indefinitely, while here is an event the like of which was never before witnessed in man's recorded history, year by year losing from its hidden store of information, and no effort is made for its acquisition.

Leonid Kulik's third and largest expedition, financed jointly by the Academy of Sciences and the Scientific Department of Sovnarkom, heads again for Siberia on February 24, 1929. It is to last a grueling twenty months, including a nightmare winter in the lonely Southern Swamp region with temperatures plummeting to a merciless minus fifty degrees Centigrade.

Here Kulik and his team of experts drill and excavate deep into the frozen craterlike pits and the massive peat formations, still hoping to find some evidence of a giant meteorite. One particularly large depression is drained by a forty-yard trench. Kulik is confident this is the object's grave. But when its base is finally cleared of dense moss, a tree stump is found firmly rooted in the center. It could not possibly have remained there if the hole had indeed been the original crater.

The team continues to work so strenuously that three of its exhausted members have to leave the expedition. Another man lies for several days unconscious and delirious from acute appendicitis before he can be moved for treatment.

Kulik's deputy is Evgenii Leonidovich Krinov, another brilliant authority on meteorites, who keeps meticulous records of the expedition. As he moves around the Southern Swamp he finds it inexplicable that, though this has been calculated as the very center of the explosion, there are parts of it with absolutely no trace of devastation. Yet, in a massive radius around it, the trees are flattened and scorched, and uprooted stumps lie where they have been blasted great distances.

On November 18 Kulik sends Krinov and another member of the team back to Vanavara with scientific samples to be forwarded to the Academy of Sciences. It is a grim journey. The rough track is now hidden beneath waist-deep snowdrifts. Krinov describes his experience in *Giant Meteorites,* one of a number of books he is to write on the Tunguska Event:

> Late at night we arrived at the expedition's winter hut on the River Chambé, our clothes wet with snow and sweat, and our leggings soaked through from crossing the taiga streams and pools. After spending the night on the Chambé, feeding the horses and drying our clothes, we continued our journey the next morning. It was a cold, frosty day and towards evening the temperature fell to 40 degrees below zero. Our leather-lined leggings, which had not dried

completely, froze, and our feet began to suffer from frostbite.

After two days, covering nearly a hundred kilometers through deep snow, we were almost completely exhausted and literally dragged ourselves to Vanavara with frozen feet. There was a great danger of gangrene setting in, so it was essential to travel at once to Kezhma for medical treatment.

After sending off urgent supplies to Kulik, Krinov and his companion spend a further six painful days traveling to Kezhma on horses obtained from some Evenki hunters at the trading station. At Kezhma, Krinov is forced to remain in hospital until mid-February, and one of his toes has to be amputated. He is not to know that vital soil samples he has risked his life to send on to the Academy of Sciences are to lie forgotten in someone's desk for twenty-eight years before being analyzed.

By the summer of 1930 the expedition's funds are exhausted, and they arrive home in the autumn. It has been a long, frustrating ordeal of sweaty, mosquito-ridden summers and unbelievably bitter winters, completely cut off from the outside world. And they have found nothing that can even begin to explain the enigma of Tunguska.

In the nineteen-thirties scientists can think of only one cosmic warhead that could conceivably have scorched and bulldozed the mighty taiga so savagely. It *has* to be a giant meteorite. Somewhere, they persist, there *has* to be a crater, or a mass of craters if it had shattered. But, despite three exhaustive expeditions, followed in 1938 by searching aerial photographs, the only relevant holes that show up are those in the meteorite theory itself. There is no crater.

Certainly the pre-impact phenomena witnessed in 1908 is consistent with the falling of meteorites. These do appear as brilliant, rapidly moving fireballs, and they *can* make terrifying sounds as their chunky masses of stone and iron power-dive into the Earth's atmosphere at anything up to forty miles a second. Like the description of the

Siberian object, they have often been likened to "fast-moving stars," and the blazing light from those seen during the daytime have been compared with that of the sun. They also have huge, luminous dust-and-gas clouds in their wake that can be hundreds of meters across. The dazzling colors of meteoric fireballs are also consistent with most eye-witnesses of the Tunguska event. According to the speed at which they are streaking by, they can vary from white, at their hottest, to blue, green, yellow, orange and finally red as they decelerate. However, this incandescence, caused by temperatures reaching some thousands of degrees, occurs with any object entering Earth's atmosphere—including our own returning space vehicles.

The kind of awesome noises described by witnesses in 1908 are also emitted by a falling meteorite as a powerful compression wave builds up ahead of it, causing a supersonic boom like that of a high-speed aircraft. Witnesses' accounts of "artillery fire and boulders falling from the sky" can be produced when a meteorite breaks up, each fragment detonating individually. Before impact there are often hissing or rustling sounds. One witness interviewed at Kansk in 1921 said he heard "something like the fluttering wings of a frightened bird."

For a human or an animal anywhere near the area of a meteorite fall the experience is quite shattering. When one meteorite dropped on England in 1965, an eyewitness compared it to a wartime dive-bomber attack and suffered severe shock for some time after.

Fireballs have been petrifying people for centuries. The following excerpt describes a meteorite fall far back in November 1662 in the Russian village of Novyre Ergi. It was recorded by a priest named Ivanishch and might well have been referring to the Tunguska Event.

> Many people saw in the clouds a terrible apparition. The sun had barely set when there suddenly arose a great star [fireball] which looked like lightning in the sky. The sky was cleaved into two parts . . . and there was indescribable light, like fire.

Afterward a small cloud appeared where the image had been. The sky seemed to close and fire fell upon the earth. Then it rose into a cloud and from this cloud poured forth noise and smoke like thunder or like a great terrible storm. For a long time the earth and houses shook and many people fell to the ground in terror. All animals crowded together, choking on their fodder, their heads raised toward the sky, bellowing each in its own fashion.

But similarities to the Tunguska phenomena no longer apply once we consider what happens *after* impact. As Ivanishch goes on: "Stones started falling, shining bright, large and small ones, all hot. They fell upon the fields and streets . . . some of the larger ones bored into the ground and froze."

More than three centuries later, in March 1976, what is claimed to be the greatest shower of meteorites in recorded history peppered 260 square miles near the outskirts of Kirin in China. Here too there is plenty of material evidence of the fall—more than a hundred pieces of stone weighing up to 1,770 kilograms that pounded their craters into the frozen soil. Witnesses describe seeing a large fireball before the meteorite exploded, flinging slabs of earth hundreds of feet into the air.

Another recent fireball, in the Dallas area of Texas in September 1971, was seen by thousands of Saturday-night patrons at drive-in movie shows. The pictures vanished in a blinding light "as bright as the sun." The meteorite, believed to have originally weighed many tons, left an ionized trail that blacked out television screens. As in other normal meteorite falls, fragments were found over a four-mile area.

There is, of course, a remote possibility that the Evenkis did find something after the catastrophe which they hid from the outside world during those twenty years before Kulik's first expedition. They were a superstitious race. Perhaps any remnants of such a heavenly body were considered sacred—some timely omen from the gods. The ancient Egyptians are known to have worshiped meteorites, and a necklace of beads made from pieces of one was found in a pyramid dating back more than five thousand years and is still preserved.

Generations of devout pilgrims have kissed the famous Black Stone meteorite at Mecca, which is regarded as holy in the Moslem world. Another found in India was anointed and decorated with flowers for many years. The Japanese revered a meteorite in one of their temples as recently as the nineteenth century, when even in the Western world few people would believe that anything larger than a hailstone could actually drop from the sky. In 1807 the trustees of the British Museum were highly dubious that four meteorites presented to them were anything more exotic than ordinary earthbound rocks.

It is easy to imagine, therefore, that among the remote and backward communities of Tunguska anything they found would have supernatural significance, especially in a land that already abounded with legends of monsters and ancient vengeful gods. The camping site of one Evenki group who lost twenty reindeer in the scorching blast was in a region of the Ognija River—a place considered accursed following an earthquake there. When the geologist Obruchev first questioned the Evenkis near Vanavara trading station in 1924, they denied the fall altogether. Obruchev wrote in the magazine *Mirovedeniye:* "In the eyes of the Tungusi people the meteorite is apparently sacred, and they carefully conceal the place where it fell."

Voznesensky of the Irkutsk Observatory also wrote in the same magazine: "The Indians of Arizona still preserve the legend that their ancestors saw a fiery chariot fall from the sky and penetrate the spot where the crater is. The present-day Tungusi people have a similar legend about a new fiery stone. This stone they stubbornly refused to show to the interested Russians who were investigating the matter."

It is extremely doubtful, however, that after so many expeditions to the devastation area over the last half century such hidden evidence would have remained undetected. It must now be accepted that whatever hit

Siberia left no trace of its origin. Never before or since has a recorded meteorite fall failed to leave solid evidence in the form of either fragments or craters. And if the Tunguska monster *had* been a meteorite it could have weighed millions of tons when it entered the atmosphere.

Certainly some of these mighty cosmic cannonballs that have been bombarding the solar system since the beginning of time are quite capable of vandalism on the Tunguska scale. Over the unruly centuries the face of the Earth has been thumped like that of a punch-drunk heavyweight—and it's got plenty of nasty scars to show for it. In fact, some scientists now believe that more than three and a half billion years ago one mighty blow virtually cracked its skull. A monumental meteorite, slightly smaller than the moon, pounded head on, they claim, smack in the middle of a vast primeval land area. The massive fracture it caused in the Earth's solid outer crust started sections of it moving apart and resulted in what we now know as the Continental Drift.

There are plenty of lesser battle scars too—like the huge Barringer Crater at Canyon Diablo in Arizona. The brute that caused it seared through the atmosphere and blasted four hundred million tons of rock out of the desert, leaving posterity to ogle at a natural amphitheater three quarters of a mile across and 570 feet deep. All but a tiny fraction of its original bulk—estimated at between 100,000 and a million tons—is believed to have been vaporized in the explosion. It has been estimated that, on average, a meteorite or asteroid capable of producing a crater of this size batters into Earth every fifty thousand years. The Arizona crater is reckoned to be fifty thousand years old. It would seem we're about due for another! But even this is dwarfed by the Bosumtwi Crater in Ghana: it is a breathtaking eight miles in diameter.

Scientist Robert Dietz, formerly an oceanographer at the U.S. Navy Electronics Laboratory in San Diego, California, has taken geological tests on a raised area of granite in South Africa called the Vredefort Ring. It is twenty-five miles across and is surrounded by a further pattern of rock folds 130 miles in diameter. He is now convinced this massive landmark was originally caused by a small prehistoric asteroid about a mile in diameter that blasted a crater ten miles deep. The asteroid, he says, could have released the explosive equivalent of a one-million-megaton bomb—enough to make the entire Earth itself shudder in its orbit round the sun. Dietz believes that the granite below the crater reacted to the removal of this great weight of earth by welling up to fill the hole and to bulge into a dome, as it now is.

Exactly where meteorites come from is a question scientists still argue over. Most are now convinced they are the airborne debris from our own solar system that frequently career off course. One theory is that they originate from the swarm of asteroids—those chunks of rock up to 480 miles in diameter that swirl between Mars and Jupiter. Astronomers in the past have suggested that the asteroids are the result of a disintegrating planet that either exploded or collided with another. One fascinating but unlikely embellishment is that some ancient race of intelligent beings that once populated this hypothetical world either blew it to bits through their own nuclear experiments or were involved in a Wellsian war of the worlds with some superpower from elsewhere in the galaxy.

The fragmented-planet idea, however, is not widely accepted. Although some fifty or sixty thousand asteroids orbit within observable range at some time, their total mass—even including the meteorites that have fallen over the years—would have been far less than even the mass of the moon. There certainly does not seem to have been enough material to form a planet of any reasonable size. The larger asteroids are now believed to have been created from leftover material in the early stages of the solar system's formation. Meteorites could be the chippings from these bodies or from the planets themselves.

A number of asteroids have orbits that reach out as

far as the Earth. One, called Eros, is twenty miles long and passed close to us in 1975. Consider what havoc a collision with that could cause.

In 1976 the American astronomer Eleanor Helin at Palomar Observatory discovered an asteroid only 12 million miles away. Astrogeologist Eugen Shoemaker says it would be "one of the easiest places to get to in our Solar System, aside from the Moon." Its Earth-crossing orbit is uncomfortably close. If it hit us at twelve miles a second it would produce a crater twenty-five miles across, even though it is the smallest asteroid ever actually measured.

Some researchers are now convinced that really super-meteorites in the distant past were responsible for many famous missing chunks of land area, like Hudson Bay and the Gulf of Mexico. The pockmarked face of the moon is certainly the surrealist handiwork of countless meteoric raids in its adolescent years. From rock samples brought back by the Apollo astronauts it now seems quite feasible that—like Earth's Continental Drift—the vast lunar depression known, ironically, as the Sea of Tranquillity might, too, have been caused by a mighty meteorite attack billions of years ago that spewed out molten rock from deep inside the moon. Work on the surveys of Mars, Mercury and Venus by Mariner space-craft show that here too there are craters up to a hundred miles wide, gouged from the desolate planets like a scooped-out Stilton cheese.

But even demolition of this magnitude cannot explain the extraordinary behavior of the phenomenal force that deformed Tunguska—a force quite unknown on our own planet. Why, for instance, do trees remain standing and virtually unscathed at the very center of the explosion? There were no natural land formations to shield them from the blast. Similarly in other parts of the cauldron, small islands of undamaged larches still stand in the midst of utter devastation.

In his book *Giant Meteorites,* Krinov writes:
The impression is that the explosive wave acted

in a very irregular fashion around the place of fall, and that it was not only the relief of the area that provided shelter. The explosive wave seems to have selected individual areas of forest for complete devastation. . . .

. . . Aerial photographs of the area round the mouth of the Churgima, i.e., three to four kilometers to the south of the Cauldron, also show clusters of preserved trees still growing. . . . A striking fact is that these trees have not only remained standing but are not even broken or scorched, although they were unprotected as a direct effect of the explosive wave. Moreover here . . . clusters of undamaged trees, dead trees still standing and those completely uprooted are found together. In these same places along the bank of the Khushmo other areas are found in which the trees, though still standing, have lost their tops, and which Kulik called "telegraph poles."

At the junction of the rivers Khushmo and Ukhagitta, twelve to fifteen miles from the Cauldron, Krinov says, "the flattened trees lie in a solid carpet, their roots sticking up into the air and facing the Cauldron, with here and there new growth showing through."

The curious way the taiga has been scorched cannot be explained, either. It bears no resemblance to that caused by any ordinary forest blaze. At the tip of every snapped-off branch is a charred cinder. In every case the fracture itself runs obliquely downward, resembling, as Kulik describes, "a bird's claw." At the top of the trees, both thick and thin branches have been guillotined in this way with cindered tips. This could have been caused only by a sudden momentary burst of tremendously high temperature. In a normal fire the slender, flimsy twigs would have been completely burned away by the kind of heat sufficient to char the tips of the thick ones.

Still faced with these baffling questions, Kulik spends a further six weeks in Siberia in the summer of 1939. Now the journey is far less hazardous. A small airstrip has been constructed near Vanavara and a thirty-five-mile road cleared through the forest leading to a campsite not far from the Southern Swamp area.

It is to be his last visit. All research is suddenly suspended indefinitely. Russia, like most of the world, now has a war on her hands. On June 22, 1941, the Germans attack. Kulik, though he is fifty-eight years old, volunteers for the Soviet Militia—the ill-trained, middle-aged "Dad's Army" in Russia's desperate all-out defense of Moscow. In October he is wounded and captured. And, on April 14, 1942, stricken with disease, the courageous Leonid Kulik, like three million other Russians, dies in a Nazi prison camp. For him, the fanatical search for the Tunguska monster is over.

4

Another "Cauldron of Hell"

It is shortly after eight o'clock on the morning of August 6, 1945. As happened over Tunguska thirty-seven years before, something is about to fall from the sky. And this is *certainly* no meteorite.

Enola Gay, an American B-29, is completing the deadliest mission in the long, tortuous history of man's inhumanity to man. Like the Tunguska object, the thing it is to release will leave no trace of itself. Neither will there be any craters. But its fury will bear the same diabolical characteristics—complete and utter devastation.

This time, however, the impact zone thirty thousand feet below is not a forest of gaunt trees in central Siberia, but a tightly packed mosaic of buildings and streets that throb with life. And the creatures that move down there are not scattered herds of reindeer, but streaming masses of men, women and children. For the 320,000 inhabitants of Hiroshima in Japan it's the start of another busy day.

The workers are already in the factories. The shops are opening up and the children are on their way to school. The Japanese early-warning radar net has spotted the aircraft above, but it's probably just a reconnaissance flight, they say. No one bothers to take cover.

The weather over Hiroshima is sultry but reasonably clear. At 8:11 A.M., Colonel Paul Tibbets, pilot of the *Enola Gay,* swings round for the run-in. He nods significantly to his bombardier, Tom Ferebe. The thing the U.S. Air Force has nicknamed "Thin Boy" plummets earthward. Inside it is uranium 235. It is the world's first atomic bomb to be detonated in anger.

In the central section of the city below a blinding white flash is all that many people see before they are blown to bits, ripped apart by flying fragments, or grilled to charcoal by the wave of searing heat that rushes out from the explosion. A dense, forbidding cloud of smoke and dust chokes the city into darkness. Through it all grope the survivors, stumbling over a carpet of corpses in blind agony and confusion.

Enola Gay swings away and heads back to base. For a while the crew say nothing. Then the tail gunner's voice whispers through the intercom, "Oh, my God . . . what have we done? What have we done?"

President Harry S. Truman is having lunch on board the United States cruiser *Augusta* when he hears that the mission is a "complete success." His prepared press release tells America that the "important army base" of Hiroshima has been atom-bombed. And for those who are not too sure what this means, the joyous statement proclaims: "It is harnessing the basic power of the universe. The force from which the sun draws its power has been loosed against those who brought war to the Far East."

Meanwhile the mushroom-shaped shroud clears from over Hiroshima to reveal the most gruesome aftermath in the annals of warfare. In a moving documentary of the rape of Hiroshima, author and researcher Robert Jungk writes in *Children of the Ashes:*

> It was no quick and total death, no heart attack of a whole city, no sudden, agonising ending that struck Hiroshima. A mercifully quick release, such as is granted even to the vilest criminals, was denied to the men, women and children of Hiroshima. They

were condemned to long-drawn-out agonies, to muti-
lation, to endless sickness. No, neither during the
first hours nor in the days that followed was
Hiroshima a silent "graveyard," filled solely with the
mute protest of the ruins, as misleading photographs
imply; rather was it the sight of movements repeated
a hundred thousand times, of a million agonies that
filled morning, noon and night with groans, screams,
whimperings, and of crowds of cripples. All who
could still run, walk, hobble or even drag themselves along the ground were searching for something,
for few drops of water, for food, for medicine, for
a doctor, for the pitiful relics of their possessions,
for shelter. Or searching for the uncountable thousands who need no longer suffer—for the dead.

Soon after the explosion, researchers like Professor
Shogo Nagaoka, a geologist from the city's university,
rummage through the mountains of ash. They come up
with some macabre specimens. The intense heat has even
melted stone and rock and twisted bottles like the handiwork of an insane glassblower. Rotting corpses shimmer
as swarms of gnats and flies cover them over with a
living shroud of black. A skinless hand, grotesquely fused
into a motorcar engine, must have been that of the
mechanic who had been repairing it when the bomb
went off.

At 11 A.M. on August 9, another atom bomb falls, on
Nagasaki. This one contains plutonium, and they call
it "Fat Boy." The target should have been the town of
Kokura, but it was too cloudy over there. Nagasaki is
the next option on the pilot's list.

The full horrific implications of these two bombs are not
to be made known until 1976, when a group of Japanese
scientists prepare a special report for the General Secretary
of the United Nations in an effort to urge the world to
outlaw nuclear weapons. These new statistics show that
the total number of deaths as a direct result of the raids
actually reaches a sickening quarter of a million. Even

now victims are dying at double the rate of the general population.

At Hiroshima, over eighty percent of people living within a kilometer of the impact center, and sixty percent of those up to two kilometers away, were slaughtered. Out of 76,000 buildings 70,000 were totally burned. And yet the force of this bomb, equivalent to twenty thousand tons of TNT, was only a fraction of that of the Tunguska monster, which is estimated to have released the shattering energy of thirty *million* tons of high explosives. Devastation in Siberia was not confined to two kilometers, but to thirty!

However, despite the vast difference in the power of the two explosions, some Russian researchers who visit Japan in the 1940s are intrigued at the many similar features. One of them is Alexander Kazantsev. He is convinced that the kind of force that razed Hiroshima and Nagasaki also flattened the forests of Siberia. For surely the atom bomb answered so many questions that had stumped Leonid Kulik in the thirties. Here in the rubble is the same kind of blast damage. The same radial scorching. There are no craters. There are no fragments of the brute that caused it. There is even the same uncanny phenomena of a few bare tree trunks still standing in the very midst of the cataclysm.

The atom bombs that were dropped over Japan didn't explode on the ground. They burst eighteen hundred feet above it. In a paper on his Tunguska expedition findings written in the 1930s, Kulik noted: "The whole central region of the blown-down woods, to judge by the remaininig dead trees, has been scorched from *above.*" The height is later estimated to have been between three and five miles.

The scorching effects at Tunguska are particularly consistent with the Japanese explosions. When the atom bombs were detonated, one third of the explosive yield was emitted as a pulse of radiant thermal energy. The heat flash lasted only a millionth of a second. And yet at ground zero, solid materials were grilled to some 3,500 degrees Centigrade. The appalling results of this massive thermal radiation were the gruesome "flash burns" that accounted for a third of the total deaths.

In 1927, even thirty-five miles away from the Tunguska impact center, witness Semenov had told Leonid Kulik, "The sky split in two and high above the forest the whole northern sky appeared to be covered with fire. A hot wind as if from a cannon blew past the huts from the north. At that moment I felt great heat as if my shirt had caught fire." Another witness had described how "a fierce heat scorched my ears."

The "heat" Semenov referred to was caused by the radiant energy from the explosion. The "hot wind" was the blast that had swept through Vanavara. But witnesses much farther away also felt the effects of the blast. A man called Bryukhanov who was 125 miles away from the impact, in a village near Kezhma, told Kulik in 1938, "When I sat down to have my breakfast beside my plow, I heard sudden bangs, as if from gunfire. My horse fell onto its knees. From the north side above the forest a flame shot up. Then I saw that the fir forest had been bent over by the wind and I thought of a hurricane. I seized hold of my plow with both hands so that it would not be carried away. The wind was so strong that it carcied off some of the soil from the surface of the ground, and then the hurricane drove a wall of water up the Angora."

Even people near Kirensk, three hundred miles away, claimed they felt a strong wind. One man wrote in a letter the day after the fall: ". . . trees were bent over and their leaves quivered. At the time the sky was clear and peaceful, the water did not show a ripple, and no damage was noticed."

There is the similarity of the mushroom clouds over Japan and that seen over Siberia. In a letter to Kulik in June 1908, the director of Kirensk Meteorological Station reported one of a number of eyewitness accounts of "a pillar of fire in the shape of a spear" and a dense cloud appearing over the spot.

Another remarkable phenomena at Hiroshima, also discovered at Tunguska, is that flowers and vegetation never seen in the city before the bomb suddenly began to flourish. Fresh lotus flowers in the ornamental pond of Hiroshima castle suddenly burst forth anew from those that had been blackened and scorched; grasses and weeds sprout everywhere. The radiation that is to kill so many more humans is a vibrant stimulus to the genetic makeup of plant life. In the Southern Swamp area of central Siberia, trees and vegetation are also reported to be growing at a phenomenal rate, with a luxuriance not known before 1908 (see Chapter 5).

The abnormally light nights that persisted over Europe and Russia following the Tunguska blast are also reported, to a less marked degree, after atom explosion tests.

All these similarities convince Kazantsev and others that it was a nuclear explosion that shattered the blue sky over Siberia. They are to be even more convinced some years later when scientists at the impact zone report excessive traces of radiation. But, as there were certainly no man-made atom bombs in 1908, Kazantsev comes to the mind-boggling conclusion that something from outside Earth must have brought it. This intriguing possibility is fully discussed in Chapter 11.

Life for most of Russia in the first few postwar years is an economic nightmare. The nation is exhausted. Its industrial machinery, its railways, its power stations—all, like its men and women, driven far beyond normal tolerance in the dogged struggle to survive—are now about as efficient as a worn-out secondhand car. Agricultural losses have been immense. As the Germans retreated they had systematically destroyed the harvests, burned down the villages and slaughtered livestock. Now food is desperately short. The 1945 harvest is a bad one. The next year's is catastrophic in an exceptionally dry season. Food prices triple and thousands of Russians starve to death.

Raw materials are now urgently needed if the Soviet Union is to survive the peace and recover its strength. For this reason Stalin gives the go-ahead for scientific exploration in ore-rich Siberia. He is also anxious to find uranium there in his anxiety to match the West in atomic power. Interest in the Tunguska mystery begins to revive, too, and a group of scientists are able to convince the government that another expedition there is worth funding.

However, at 10:38 on the morning of February 12, 1947, another monster from outer space screams across Siberia. The first spectacular characteristics appear the same. Thousands see the dazzling fireball, the brilliant fiery tail spitting out sparks like a rocket, the great track of swirling smoke curling from horizon to zenith that hovers for several hours after the object pounds into the Sikhote-Alin mountain range near the Chinese border. Again there are those same fearful sounds: the thunderous rumblings, the rapid cracking like artillery fire.

Listening to eyewitnesses is like listening to the stories that the Evenki people told to Kulik more than twenty years before. One man, N. F. Kushnarev, who was only four and a half miles south of the fall, tells of a "great flash, brighter than the sun," that hurt his eyes. He then saw a dense black streak of smoke in the sky, followed by an explosion that caused houses to tremble and a series of crackles "like a machine-gun salvo." Others talk of doors being blown open, windows blasted out and plaster showering down from ceilings. Two pilots, Firtsikov and Ageev, at a nearby airfield, describe the fireball as being "as large as the moon" as it disappeared over the hills. Horses and cows whinnied and lowed in terror, broke free and bolted about in all directions. Dogs howled and fled into the forest. A mechanic, V. I. Esteev, who was working on a telephone pole at the time, felt a severe electric shock from the wires, even though the line was disconnected.

After the first reports the investigators waste no time in scurrying to the fall area. Here, at last, could be the elusive answer to the riddle of Tunguska. A hurriedly assembled team of geologists is dispatched by air. But, instead of solving the riddle, the expedition findings merely make it even more confusing. For here at Sikhote-Alin the evidence of a normal meteorite bombardment is obvious. There are 122 rusty-colored craters, some up to thirty feet deep and eighty feet wide, pockmarking the snow. Meteoric iron is abundant on the mountain slopes and lying among the rocks. But there are no traces of blast and scorching, none of the explosive phenomena found in 1908. In fact, the meteorite's impact was not even registered by the sensitive instruments of the nearest seismic station, three hundred miles away at Vladivostok. It all strengthens the view that the Tunguska invader was no ordinary meteorite.

Unfortunately the Sikhote-Alin fall is to dominate meteorite research by the Academy of Sciences for the next three years, during which time the academy sends four expeditions there. Again the Tunguska mystery has to be shelved.

However, in 1957 the specimens of soil brought back by Leonid Kulik, which had lain forgotten at the academy for more than a quarter of a century, are finally unearthed and analyzed with the more sophisticated apparatus now available. Some samples appear to contain tiny iron-dust fragments, possibly meteoric in origin. The academy therefore launches its largest expedition to Tunguska to settle the matter once and for all.

The expedition is equipped with the very latest instruments and aerial maps, and its leader is Professor Kirill P. Florensky, one of Russia's most respected geochemists, who flew over the impact region in 1954. For Florensky and his scientific team in June 1958, there are to be none of the hardships Leonid Kulik had been forced to endure. No crippling weeks driving primitive horse-drawn wagons over hundreds of treacherous miles of uncharted country. For Florensky, getting to Vanavara means merely a few hours' flight on a comfortable aircraft from Kezhma. The derelict little trading station of

Kulik's time is now a bustling center in full radio contact with the outside world. It has its own airport, hospital and school. Neither is there to be the tortuous struggle against the floods and rapids of the river journey into the taiga. Instead fast and powerful motorboats are waiting to transport the tons of valuable new equipment.

And yet, after thirty-four days in which they search five hundred miles of forest, the team still finds no sign of meteorite craters or fragments. In fact, it discovers that those small iron "meteoric" particles analyzed in Moscow before the expedition left are artificial and of terrestrial origin. They came from various metal instruments such as bores and spades that had been used in the earliest expeditions.

Part Two

5

The Monster That Makes Things Grow

By the late nineteen-fifties, despite five major expeditions and dramatic development in all fields of technology, the world's scientists are still unable to uncover the secret of Tunguska. It is now half a century since the monster struck. Vast new forest growth veils the whole area, just as age veils the memory of those original eyewitnesses still alive.

Basically all that the researchers can confidently state is that nothing else like it has ever been recorded; that whatever it was pulverized itself in midair and did not strike the ground; and that the only known force approaching its ferociousness is that of a nuclear explosion. The radiation from a nuclear explosion, too, could account for the incredibly high mutation rates being reported in Tunguska's Southern Swamp.

In 1959 Dr. Gennady Plekhanov, of the Betatron Laboratory of the Tomsk Medical Institute, tests three hundred soil samples and one hundred plants from the area. He claims that his analyses clearly show that "in the center of the catastrophe radioactivity is one and a half to two times higher than that found thirty to forty kilometers away." Other analyses show the level to be six times above normal.

Researchers are finding that since 1908 trees in the impact region have grown at a fantastic rate. Those that have sprouted since that date are twice the normal height and their trunks three times as thick. Some plant growth is reported to have been stepped up twelvefold.

Examinations of the growth rings in older trees shows that from the time of the explosion cell production has increased dramatically, causing the rings to be broader. Traces of the radioactive isotope cesium 137 is also claimed by some researchers to be present in the ring structures.

Even in 1931, Kulik had remarked on "the surface of the swamp being luxuriantly overgrown with sphagnum [bog moss] whose age does not exceed twenty years." And one Evenki had told him, "How overgrown it has become. It was not like this before." It appears that the high radiation level in the swamp area is acting as a stimulus to all forms of vegetation growth.

In 1975 a series of scientific papers concluded that there was also a mutant effect in the tree population along the trajectory taken by the Tunguska object. Edited by academician V. S. Sobolev, leading authority on current research into the Tunguska Event, the papers were published by the Meteorite and Cosmic Dust Commission of the Academy of Sciences in a book whose title, *The Problematic Meteorite,* emphasizes the mystery that still surrounds the explosion.

Though no evidence exists of any mutations affecting the animal life there, these could conceivably have taken place. Kulik's earlier expeditions would not have been looking for such abnormalities, and by now these could have been obliterated in the normal course of natural evolution. In most animal groups, weak or deformed members do not normally live for long. If they cannot feed themselves they starve to death. If they are a burden on the rest of the group they are killed. There is little chance of them reproducing further mutants. In the case of a deformed male reindeer, for example, it would most likely be killed off by its competitors during the mating season. Nature is extremely selective. Only the fittest survive.

Another probability is that if any malformed animals had roamed the taiga the Evenkis would have slaughtered them. They were a highly superstitious people. Such strange afflictions in their herds could well have been interpreted as bad omens. They would have been reluctant to tell strangers about them for fear of aggravating the gods.

However, the large-scale mutation in the vegetation could eventually lead to changing animal species as their feeding habits become adapted to the new varieties of growth. This would, of course, take many generations.

There have certainly been no reported cases of human beings suffering from mutation effects, though here again, if any child *had* been born with a deformity, those same primitive beliefs might have induced the Evenkis to hide the offspring.

Let's consider briefly how radiation causes mutation. Our physical characteristics are dependent on the genetic coding in our cellular makeup. This coding is in turn passed on to our children. If, for any reason, this pattern is broken, a genetic change takes place. This can be brought about by severe chemical stimulus. But the most rapid cause is for part of the genetic structure to be destroyed by radiation. If the exposure is particularly drastic, the individual's characteristics will be affected. It has been found that fruit flies exposed to radiation rapidly breed others with such mutations as deformed legs and no wings.

Recently Britain's National Radiological Protection Board and the Department of Health began a two-year nationwide survey to find if medical X rays were causing genetic harm. One of the potential dangers of X rays is that they can damage chromosomes which carry the genetic code. Doctors are warned not to X-ray pregnant women. Dr. Stewart Rae, assistant director of the board, is quoted as saying, "We want to find out just what sort of genetic burden we are imposing on future generations."

Apart from radioactivity, soil tests taken at Tunguska in the late fifties also reveal strange globules composed of magnetite and silicate. They are either spherical or in the form of droplets measuring only a few thousandths of a millimeter across. Some are hollow and completely transparent, and one striking feature is that some of these "cosmic ball bearings" are coalesced or fused into solid masses—an indication that they must have been subjected to tremendous heat. (See Appendix I.)

Krinov sees all this as more evidence for the comet theory. He writes:

> Meteoric bodies when moving through Earth's atmosphere are destroyed by fragmentation and by the melting of their surface layers. The molten layer is blown off by the air currents encountered and sends out a spray of minute droplets that are scattered in the atmosphere. Because of their small size these droplets harden to form tiny globules. Such globules form the dust trails which are observed for some time after the bolide [fireball] flashes across the sky and a meteorite falls. Research has proved that when meteoric bodies composed of iron are destroyed, they form globules of magnetite, i.e., the globules differ in composition from the meteorite body from which they originated, due to oxidization of the molten droplets of iron by the air.
>
> When meteorite bodies of the achondrite or the chladnite types, i.e., consisting almost entirely of the snow-white mineral enstatite, and with practically no intrusion of nickeliferous iron, are destroyed, transparent or white silicate [glass] globules are formed. . . .
>
> A whole series of facts—the appearance of an anomolous optical phenomenon in the atmosphere after the fall, the high initial speed of the meteoric body, the explosion in the atmosphere, and, finally, the discovery of the globules of different composition—lead to the conclusion that the Tunguska meteorite was a comet.

But there are still unanswered questions. Russian analysts now claim that the particles from Tunguska differ from any found in normal meteorites. The iron traces there are not substantially higher than the normal accumulation of meteoric dust from the sky that is slowly deposited all over the world. Even Krinov admits *"it will take many years before the remarkable phenomena of Tunguska can be regarded as fully investigated and explained."*

The nearest natural explanation that anyone comes up with in the nineteen-fifties is the comet theory. It is suggested that the nucleus of a very small comet plunged into the atmosphere at such a high speed that the heat generated caused it to explode. Already American astrophysicist Fred Whipple had suggested that the pulverized tail of a comet caused the phenomena of the bright nights recorded in many parts of the world. The tail, directed away from the sun (then in the southeast) would have stretched over Europe, where it was captured by the upper layers of the atmosphere.

A number of eyewitness reports were consistent with comets. In 1935 a letter from one eyewitness to Leonid Kulik describes the Tunguska fireball as a "brilliant white mass in the form of a ball of cloud with a diameter far greater than the moon." On July 1, 1770, Lexell's Comet, passing closer to Earth than any other is known to have done before or since, appeared over two degrees, thirty minutes in diameter—five times the apparent diameter of the moon. And many centuries before the fireball of 1908 terrified the Evenki people observers were describing comets in similar terms. One report, recorded on a stone tablet in Babylon and dated 1140 B.C., reads: ". . . a comet arose whose body was bright like the day, while from its luminous body a tail extended like the sting of a scorpion." Another record, from Syria in 146 B.C., describes "a comet as large as the sun. Its disc was at first red and like fire, spreading sufficient light to dissipate the darkness of the night."

Comets can vary right through the range of colors reported by witnesses in Siberia. In forty-nine sightings recorded by the Chinese, twenty-three were white, twenty bluish, four red or reddish yellow and two greenish. However, no description of the Evenkis could surely match this lurid one by Ambroise Paré, a distinguished French surgeon of the sixteenth century: "This comet was so horrible and so frightful and it produced such terror in the vulgar that some died of fear and others fell sick. It appeared to be of excessive length and was the color of blood. At the summit of it was seen the figure of a bent arm, holding in its hand a great sword, as if about to strike. On both sides of the rays of this comet were seen a great number of axes, knives and blood-colored swords among which were a great number of hideous human faces with beards and bristling hair."

Comets have been regarded with dread and suspicion by most of our ancestors. They have been interpreted as celestial messengers of the most unsavory warnings, like death, destruction and disease, war and famine, the dethronement of monarchs, the fall of empires, and even the end of the world. They were said to have predicted such shattering events as Nero's death, the Great Plague of 1665, the Great Fire of London and the dividing of the Red Sea during the Israelites' exodus from Egypt. The arrival of Halley's Comet has been associated with a number of historical highlights. In 1066 it was the Norman Conquest, and in 1696 a church minister suggested it had caused Noah's Flood. When it returned in 1910 some people in the Middle East are reported to have committed suicide. It is due to visit us again in 1985.

So what exactly is a comet? The remarkable thing is that even today no one is quite sure, even though some three hundred new comets are identified every century and it has been estimated that there could be as many as five million in existence. We know that they travel in orbits round the sun, just like the solar planets, and occasionally pass very close to Earth. We know that some of them appear at regular intervals ranging from a few years to many centuries, depending on how long it takes them to encircle the sun. We know they consist of a bright, starlike center, or nucleus, surrounded by a cloudy head (the coma) which blends into a long, luminous tail that can be one of the most spectacular sights in the sky. But precisely what the nucleus is made of is still debatable. There are three principle theories:

1. That it is loosely packed ice and frozen gases like ammonia, methane and carbon dioxide in which are embedded minute particles of solid matter. Recent spectroscopic analysis shows that sodium, nickel, chromium, silicon, calcium and magnesium can also be present.

2. That it is made up of large chunks of solid matter, some up to a mile or more in diameter, and is surrounded by gas and dust. The famous Kahoutek Comet, last seen in 1973, is thought to have been composed of a tough, stony silicate outer surface protecting volatile gases inside.

3. That the nucleus and the surrounding coma are a great cloud of gas molecules and dust particles.

Whatever it is made of, its actual mass seems to be very small. For although the entire nucleus can be thousands of miles across, its average density is believed to be a mere half an ounce per cubic mile, compared with five million tons of air on Earth.

A great deal of research is now being done to find out the chemical and physical processes that occur in comets. In 1970 the Orbiting Astronomical Observatory (OAO 2) was able to study the comet Tago-Sato-Kosaka from beyond Earth's atmosphere. Using an ultraviolet telescope, it found that the comet itself was enveloped in a huge cloud of hydrogen vaster than the diameter of the sun. The same year, the Orbiting Geophysical Observatory (OGO 5) found that the brilliant Comet Bennet was surrounded by a similar cloud an enormous eight million miles across. But even these were long-distance observations. Now scientists hope to dispatch space probes to fly by a comet—or even to attempt a sensational soft landing on the actual nucleus. The return of Halley's Comet could be a heaven-sent opportunity.

No one really knows where comets come from, either. The most common theories are the following.

1. That they are formed from gaseous matter ejected from the sun which somehow escapes into space and is prevented from falling back to the sun by the gravitational pull of the giant solar planets.

2. That they are material from the giant planets themselves, perhaps blasted into space from huge volcanic eruptions.

3. That they are bits of a planet that was once part of the solar system and disintegrated in some catastrophic explosion or collision—like the chunky belt of asteroids that orbit the sun between Mars and Jupiter.

4. That they are the leftover debris from the time the solar system was first formed.

5. That they are from a huge belt of frozen materials called Oort's Ring that orbits the sun at eighteen billion miles.

When a comet's material begins to break up, tiny fragments of it splay off into their own slightly different orbits, forming a belt along the comet's path. If the earth passes through them, they can enter its atmosphere to produce meteor showers (shooting stars). Those that don't entirely burn away in the process arrive here as solid meteorites. We are well aware of the havoc that even a giant meteorite can cause. But suppose the entire nucleus of a comet careered off course and pounded to Earth. Perhaps its lethal potential might even be as awesome as that calculated by the French mathematician Maupertuis, back in the eighteenth century, who wrote: "There are some comets so small that collisions with Earth would destroy only a few kingdoms without shattering its mass, but there are others the contact of which might be fatal to every living thing on the globe."

The comet theory, incidentally, does not rule out the possibility of the Tunguska devastation having been caused by a nuclear blast. All such explosions need not be man-made.

Let us assume that in 1908 a comet nucleus with a stony outer shell, like that of Kahoutek, enters Earth's atmosphere at something like twenty-five miles a second. The only heating it has experienced on its long journey through free space has been that from the normal radiation of the sun. Now it suddenly begins to glow at far greater temperatures, enough to begin stripping away the stony protective layer until the highly volatile gases inside—originally frozen—heat up drastically and try to escape through the stony shell. But they are held back by the brute force of the shock wave that has formed in front of the comet. Finally, however, they produce an almighty armor-piercing blast which bursts through the shock wave. This process immediately heats up the gases, largely hydrogen, to the critical limits of temperature and pressure when they are transformed into helium. What we now have is the basis for a natural nuclear "bomb" explosion—the kind that takes place continuously inside the sun itself.*

Such a thermonuclear reaction could conceivably occur a few miles above the ground, which might cause the kind of devastation found around the Southern Swamp and also be consistent with the blast pattern at Hiroshima. However, if the object *was* a comet, it is the only recorded instance of one reaching Earth, despite the vast number that streak through the solar system. The closest any is known to have approached in modern times is three and a half million miles.

But there are many other cosmic forces even more inexplicable and ferocious than comets and meteorites. And so, while men like professor Florensky are still desperately trying to dig up the past from the mud of Tunguska's Southern Swamp, others are looking far beyond Earth itself for the answers. That's where the mon-

---

* A fuller, revised version of an exploding comet head is given in Appendix II.

ster came from. Surely that's the place to look. For somewhere out there could be its celestial breeding ground. Somewhere out there could be many others of its kind.

It is a good time for such a search. Man is taking the first timid steps into space. Already, in 1957, he has sent an inquisitive little satellite called Sputnik I scurrying round the globe. For the first time everyone *knows* there is life in space—even though it's only a Russian dog. Four years later man tries it himself. By 1969 two Americans are actually leaving footprints in the lunar dust. By the seventies satellites are busy monitoring the remotest corners of the world while spaceships infiltrate the solar system and even land robots on the planets.

From it all is emerging a fascinating and far-reaching new technology. No longer is man confined to digging in his own terrestrial back yard for the relics of the facts of life. Now space research is giving him a fresh and exciting vision into not only his own creation but that of the universe itself.

If any fading vestiges of the thing that created the Cauldron of Hell still remains, what hope is there of exhuming it now after the searching ravages of atmospheric decay, the churning geological unrest and the ever-changing shroud of prolific vegetation? On the other hand, though seventy years may be an earthly life span, in astronomical terms they compose but a fleeting moment in time. Out of erosive reach of those aging, withering influences of our own environment, life can be sterilized for eternity and gauged not in decades but in billions of years.

And so we shall now continue to investigate possible answers to Tunguska far beyond the taiga of central Siberia—even beyond our own solar system. For among the strange and inexplicable predators that lurk in this endless cosmic jungle could well be the counterpart of the thing that left its abrupt reminder in 1908 that this Earth of ours is far from invulnerable.

6

Fury of the Black Hole

Could the unprecedented force that lashed Tunguska and vibrated round the Earth in 1908 have been compressed into a body the size of a speck of dust?

Inconceivable though this theory may seem, it has been put forward by two highly qualified scientists from the Center for Relativity Theory at the University of Texas. And in September 1973 their findings were published in *Nature,* one of the world's most respected and fastidious scientific journals. The object, they say, was a "black hole." (See Appendix III.)

In considering such a possibility we must now smash through all normal barriers of credibility. We must peer far beyond the limits of our own logic, conception and experience. A black hole in the Cauldron of Hell! Could anything sound more forbidding? But remember, we are now infringing on a new and bizarre region of space research where science fact is emerging as even stranger than science fiction.

We are also attempting to unravel a happening that every known and tested avenue of man's scientific scrutiny has failed to classify. As Krinov wrote in 1966: "The effects associated with the immense explosion which occurred on June 30, 1908, are unlike those reported for any other natural or man-made phenomenon." And an assessment of the black-hole theory published in the Royal Astronomical Society journal in November 1975 states: "The apparent uniqueness of the event requires that all possible explanations must be seriously considered and no explanation can be discarded merely because it has a low possibility of occurring."

A black hole is virtually the celestial phantom of a star originally many times bigger than our own sun. It aquires this macabre state when its matter collapses in on itself and shrinks beyond a point of no return. In condensing to merely a few miles in diameter, its phenomenal power of gravity remains to dominate every other force. It devours its own body and everything about it. Nothing we know of appears able to withstand its frightening ferocity.

We can't see black holes because even the light from their own tiny white-hot nucleii is dragged back before it can escape the supergravity that envelops it. We know they exist only because they have been found to affect even the movement of mighty stars in their vicinity.

It is now believed that gigantic black holes could be at the centers of entire galaxies of stars, causing them to gather in clusters as they do; such is their overwhelming power. In April 1976, researchers at the Cambridge, England, Institute of Astronomy suggested that a black hole with the mass of ten million suns could be the source of intense X rays being picked up from the galaxy known as Centaurus A. The nucleus of this galaxy emits nearly as much X-ray power as do all the 150 billion stars in the Milky Way (our own galaxy). It is thought that this black hole was originally a supermassive star or a dense cluster of stars that is now compacted within a radius less than fifty times that of the sun.

Actually the X rays would be originating from stray matter which the black hole is drawing into itself like an enormous vacuum cleaner, and which is being heated to temperatures of hundreds of millions of degrees. A black hole of this dimension could be systematically devouring nearby stars that venture too close. For, like any other predator, the black hole must go on seeking new prey in order to survive. In astronomical terms its life expectancy is not great. If it doesn't ingest sufficient material, it literally starves to death.

Cygnus X, an insatiable black hole discovered in our own galaxy, is sucking huge clouds of hot gas from the atmosphere of a large, visible star around which Cygnus X is in orbit. Eventually the entire star could be swallowed up into the black hole, leaving no trace of its former existence. Consider, then, the appalling result of our own Earth finding itself anywhere within clutching distance of such an horrific force. Nothing could stop it from being ruthlessly torn apart.

As an example of how this gravity buildup takes place, imagine that our Earth, which has a gravitational pull of 1 g., could be reduced to a quarter of its present size (i.e., a diameter of only 2,000 miles), but somehow retain its present mass. The gravity field on our highly compressed planet would then become 4 g. If the diameter was reduced even further, to 500 miles, the inward pull would be 16 g.—a force that would crush the human frame. Based on this fixed-square-law relationship between the size of a body, its mass and the gravitational field it exerts, by the time the Earth was compressed to a mere mile in diameter the force field would be an incredible 64 million g.

After such massive shrinkage the basic characteristics of the very atoms of which the planet is composed would collapse. To compress Earth beyond this point would produce a critical limit at which no known structure could support the force field produced. Nothing could then stop the collapse to a pinpoint. The planet would then be a small black hole. A similar object hitting our world would destroy it as utterly as if we had collided head on with Venus. And the tiny, invisible vandal would merely continue on its way.

To produce merely the kind of energy release that

violated Tunguska would require a black hole tinier than a speck of dust. Yet it would be so dense it would weigh an astounding ten thousand million tons!

A. A. Jackson and Michael P. Ryan of Texas University suggest that such an object hit Siberia, drilled its way right through the Earth and emerged from a point in the North Atlantic.

> We suggest that a Black Hole of substellar mass . . . could explain many of the mysteries associated with the event [they write]. The Tunguska meteorite left a visible fiery trail accompanied by thermal radiation and a blast wave that levelled forest over several hundred square kilometres. No crater and no meteoric material that can unambiguously be associated with the event have ever been found.

These characteristics, they claim, are consistent with the impact of a Black Hole entering the surface of the Earth. Such an object approaching our planet would draw in constituents of the atmosphere from a wide area around itself. Its temperature would rise to between 30,000 and 300,000 degrees Centigrade, and the ultraviolet radiation would produce a plasma column that would appear as deep blue in color.

"These results," continue Jackson and Ryan,

> agree well with eye-witness reports of the event and with measurements of the pattern of throwdown of trees at the site. Eye-witness reports of flash burning, descriptions of the object as a bright blue "tube" and searing of trees at the site indicate a compatible temperature.
>
> Since the Black Hole would leave no crater or material residue, it explains the mystery of the Tunguska Event. It would enter the Earth and the rigidity of rock would allow no underground shock wave. Because of its high velocity and because it loses only a small fraction of its energy in passing through the Earth, the Black Hole should very nearly follow a straight line through the Earth, entering at

30 degrees to the horizon and leaving through the North Atlantic in the region of 40 to 50 degrees N, 30 to 40 degrees W.

However, there are a number of holes (other than black ones) in the Jackson and Ryan theory. Firstly, their geographical coordinates mean that the object, after melting a tunnel through the rocks of the Earth, should have emerged to the northeast of the deep Atlantic Ridge, roughly in an area known as the Flemish Cap. From here, the nearest land area is St. John's, Newfoundland, approximately eight hundred miles away. About fifteen hundred miles to the east is the busy port of Lisbon on the Iberian Peninsula. Here the enormous tidal wave such a disturbance must have caused would most certainly have been recorded, even in 1908. We are dealing with a concentrated energy release up to one hundred thousand times greater than that of the largest known earthquakes. And even these can produce tidal waves of thirty feet. As far as is known no exceptional ones were reported at the time of the Tunguska Event.

A force such as that of even the minutest black hole passing so close to the North Atlantic Ridge, a major crustal scar in the seabed, could have torn open the "wound," sending a massive split along the ridge, causing colossal volcanic eruptions. Such a cataclysm could not possibly go unnoticed.

Jackson and Ryan suggested that microbarograph records should be checked for evidence of shock waves from the emerging black hole. Two other researchers at Texas University, William Beasley and Brian Tinsley, did this, and their findings were published in *Nature* in August 1974. They estimated that the object would have taken ten to fifteen minutes to pass through the Earth, causing an explosion at the place of exit similar to the one that occurred at Tunguska. They point out that it took the Siberian waves five hours to reach meteorological stations in London. So similar waves from the North Atlantic should have arrived there, too—some three hours earlier.

"We have examined copies of the English microbarograph records," they say, "but have been unable to find any sign of waves from the suggested exit explosion." Not only should these have been recorded, but they should also have been stronger. They would not have had so far to travel.

Further objections to the black-hole theory are also contained in a paper published in the Royal Astronomical Society journal in 1975. The authors are Jack Burns, George Greenstein and Kenneth Verosub of the University of Indiana, Amherst College and the University of California. They claim that the point where the black hole entered the Earth "should be marked by a patch of melted and resolidified rock of diameter half to four kilometres, overlain by fused soil of comparable extent. As the hole entered the soil it would have vaporised the water, oxidised the organic matter and fused the residual material such as quartz, feldspar and mica. . . . The point of impact should therefore be marked by a depression."

But Burns, Greenstein and Verosub have to agree that there *is* in fact a depression at Tunguska: the Southern Swamp. However, they add, "This depression may predate the Tunguska Event and is not inconsistent with other explanations."

Once inside the Earth, a black hole would be expected to cause drastic stress changes in the surrounding rock, resulting in large-scale earthquakes. But the seismograms that registered the Siberian blast showed it to be consistent with a *surface* wave and not with extraordinarily violent shocks underground.

However, until we know far more about the behavior of these supercharged monstrosities from outer space—and many scientists now believe them to be scattered in vast numbers throughout the universe—it is difficult to entirely rule out the possibility, however remote, that one of them was just "passing through" on a visit here in 1908. There is not a scrap of recorded evidence showing exactly what such an impact would cause. They may have been here before. Like the detonations of comet heads or the fall of gigantic meteorites, black holes too might well have sculptured the very face of our planet in its early history.

7

Antimatter:
The Ultimate Force

During the early thirties, when Leonid Kulik was still firmly convinced that the Cauldron of Hell was the brutal handiwork of a meteorite, another scientist made an astounding discovery that could yet prove him right. For, according to the remarkable findings of P. A. M. Dirac, the brilliant British physicist and Nobel Prize winner, there could be a form of meteorite infinitely more fearful than the kind for which Kulik was searching—one that could have produced most of the Tunguska phenomena and still left no crater or any telltale fragments of itself. It could also have accounted for those unusual flash burns. And it would not have had to weigh a million tons. Or even one ton. It need have been no larger than a cube of sugar. For this meteorite could have been composed of the most potentially violent explosive material in the universe—*antimatter.*

The moment antimatter encounters ordinary matter there is complete annihilation. The energy released is greater than that from any other source known to science. Six cubic inches of antimatter pounding into our Earth could produce a blast capable of flattening the British Isles. Quite simply, this remarkable material from the exotic world of high-energy physics is composed of atoms identical in every respect to those in ordinary material—except that everything is constructed in reverse.

To appreciate this, first consider the makeup of normal atoms, of which everything on Earth, including ourselves, is formed. An atom has a positively charged nucleus called a proton. Around it, one or a number of negatively charged particles called electrons spin in dizzy orbits at the inconceivable rate of 100 million billion circuits a second. The opposing charges cancel each other out so that the entire atom itself has *zero charge*.

A similar atom of antimatter would be constructed in precisely the same way. It would be the same size, the same weight. But, in this case, the electrical charges would be reversed. In other words, the central proton would have the negative charge, and the "satellite" electron would have the positive charge. One atom, in fact, would be a mirror image of the other. (See Appendix IV.)

If these two atoms meet, the positive charge of the ordinary proton annihilates the negative charge of the antiproton. At the same time the negative charge of the ordinary electron is annihilated by the positive charge of the anti-electron (known as a positron). The net result is that they vanish in a tiny but intense flash of gamma radiation. This belligerent alliance produces the highest known energy yield.

Most people are vaguely aware that today's physicists are able to tap the enormous energy trapped in the whirling maelstrom of electrons and nuclei that form the atoms of normal matter. They see the results in nuclear-driven submarines, aircraft carriers, icebreakers and tankers. So awesome is the source of these vessels' power that the tankers have been barred from entering some of the world's harbors.

But just how do we obtain and harness this basic energy of the atom? What *is* nuclear power? What makes an atom or thermonuclear bomb? And what are "fission" and "fusion"?

The power within an atom lies purely and simply in

its mass. If we cause this to change, we obtain *energy.* This stems directly from Einstein's famous equation: $E = m\,c^2$ (where E is called "rest mass energy," m is the mass of the material and c is the velocity of light).

What this means in layman's terms is this: *If we destroy the mass, we get pure energy in the form of radiant light.* And there is only one way of *totally destroying* a mass of matter (whether it's iron, lead, water or anything else), and that is by combining it with an equal mass of antimatter. The result is that tremendous blast of gamma rays—a form of "light" that we cannot see, but that is capable of penetrating solid objects which are even opaque to X rays. This is called *annihilation energy.*

Up to the present time scientists have discovered no way of safely harnessing such total energy. But they *are* able to produce a fraction of it through "fission" and "fusion." Nuclear *fission* is the energy obtained by the splitting up of large heavyweight atoms. Nuclear *fusion* is the combining of very light atoms to form heavier ones. We know the catastrophic results of these processes in the A-bomb and its more powerful successor, the H-bomb. Yet these are infinitely less lethal than the total annihilation energy produced from mixing matter with antimatter. (See Appendix V.)

The energy release in an A-bomb is produced by the loss of only 0.4 percent of its original mass. A startling example of this is the Hiroshima bomb, whose initial total explosive size was roughly that of a standard can of baked beans. The actual quantity of matter consumed to *produce* the explosion that devastated the city would have been about the size of a grain of birdseed!

Scientists can observe the behavior of antimatter particles by sophisticated laboratory techniques. But they require colossal amounts of energy and extremely powerful instruments to produce them in even the most minute quantities. Even then the antimatter particles survive only until they combine with normal particles—a matter of a millionth of a second. Only the complicated equipment of the high-energy physicist can record the traces of radiation that are produced.

What is more mind-boggling is that for every particle of ordinary matter there is believed to be an antimatter counterpart. This means that there could be regions in the unpredictable realms of outer space where *everything* is composed of antimatter—even entire galaxies, populated by antimatter stars and planets. There is no way we could distinguish them from normal heavenly bodies. The light generated by the stars would be exactly the same as that from any in our own galaxy. If any of these planets were inhabited by intelligent life, that, like everything else around it, would be fashioned in antimatter. They would be quite unaware that any other kind of material existed. Unless, of course, they happened to bump into *us*. In that case the first historic meeting between man and "anti-man" would produce the most devastating handshake in the history of community relationship.

Research is still going on into the intriguing possibility that antimatter did, in fact, cause the Siberian blast when a mass of it exploded soon after entering the Earth's atmosphere. There are a number of consistencies. The brilliant blue streak described by many eyewitnesses in 1908 is precisely the visual effect that a chunk of antimatter would produce as it drove toward Earth and began to react with small scraps of ordinary material in the atmosphere. The nearer it got to us, the denser would be the atmosphere and the more violent the reaction.

Though actual impact with Earth would seem the most likely cause of a final full-scale explosion, it is conceivable that a powerful shock wave occurred in front of the falling body and predetonated it, resulting in a midair blast. This would cause the same radical destruction and flash burns found at Tunguska.

Williard F. Libby, ex-member of the Atomic Energy Commission whose discovery of radioactive-carbon dating won him a Nobel Prize, has suggested that the nuclear processes resulting from the annihilation of a meteorite of antimatter would have increased the content of carbon 14 in the atmosphere. Had this happened, he said, trees all over the world soon after the explosion should have absorbed abnormally large traces of it. The carbon-14 content was therefore measured in the rings of a three-centuries-old fir tree in Arizona and an ancient oak near Los Angeles. Remarkably, the highest levels were found to have been absorbed in 1909—*the year following the Tunguska Event*. However, this evidence is not accepted as conclusive. Carbon 14 can build up through normal processes. It is produced at a fairly regular rate by the bombardment of cosmic rays from space. Libby believes that a meteorite entirely composed of antimatter would have resulted in even bigger increases in carbon 14 than those actually detected.

Dr. Hall Crannell of the Catholic University of America in Washington, writing in *Nature* in 1974, suggested there could be other materials whose presence at Tunguska in any considerable quantities would indicate a matter–antimatter explosion having taken place. One of them is the radioactive isotope aluminum 26, which could be traced by counters. These tests have not so far been carried out. (See Appendix VI.)

There is one obvious loophole in the antimatter-meteorite theory as applied to Tunguska. It is unlikely that many of these frightening little cosmic warheads retain their lethal power for long once they encounter the ordinary interstellar matter of a galaxy such as our own. They certainly couldn't have originated in the Milky Way, which means they would have had to endure incredibly long journeys over many billions of years. Even if some of them began their marathon flights as gigantic hunks of rock, they would inevitably be whittled down by countless minor explosive impacts with interstellar particles of ordinary matter on the way. And the sniping would be even more concentrated once they streaked into the dusty, gas-laden regions of the solar system itself. In addition, the objects would have to negotiate an endless gauntlet of various suns and planets long before they got here, and somehow avoid being dragged toward them by the individual forces of gravity.

However, like the black hole, the antimatter meteorite cannot be entirely ruled out as the Tunguska monster. Here again there are so many unknown factors, so many questions our scientists must still answer in an unfamiliar field of research of which, as yet, we are only on the fringe. Antimatter may have somehow found ways through to us many times in the past. There are other unaccountable depressions dotted about the world similar to the Southern Swamp of Tunguska.

One intriguing theory is that ball lightning—that strange will-o'-the-wisp phenomenon frequently witnessed all over the world—could be caused by tiny fragments of meteoric antimatter from the upper atmosphere. (See Appendix III.)

David Ashby and Colin Whitehead of the Atomic Energy Research Establishment at Harwell, England, suggest that a "hypothetical barrier" exists between matter and antimatter. A minute particle of antimatter might, therefore, be comparatively stable if it was traveling at a low speed, whereby colliding air molecules would be unable to get through this "barrier." During thunderstorms they may aquire enough energy to annihilate with ordinary matter, resulting in an energy release that would form a luminous ball. Ashby and Whitehead have actually observed several short bursts of gamma rays, a phenomenon also associated with matter–antimatter annihilation.

About seventy percent of ball-lightning outbreaks appear to occur during periods of thunderstorms, and these are particularly prevalent and violent in Siberia during the summer months.

Some years ago the Soviet authorities were baffled by frequent and unaccountable disappearances of aircraft on long-distance flights from Vladivostok to Moscow. Often the last radio messages from aircrews described "brilliantly glowing balls" moving toward them. Days later the mangled wreckage of the aircraft would be found scattered over some desolate area. There were never any survivors to tell exactly what had happened. It was Academician Peter Kapitsa, at one time head of the Academy of Sciences and a specialist in high-energy electrical discharges, who insisted that the disasters were caused by ball lightning. This eerie phenomenon takes the form of rapidly spinning balls that glow vivid blue, violet, pink or green according to the spectra of the dust that is contained in them, skipping briskly about the sky like Peter Pan's fairy Tinkerbell—but infinitely more lethal. Without warning, and within seconds or minutes of appearing, they often explode violently. If one of these little horrors forms inside a building, it can cause extensive damage.

Kapitsa actually produced ball lightning in the high-voltage laboratories of Moscow University. It was almost a Dr. Frankenstein operation. To simulate even the tiniest specimens, he had to momentarily short-circuit the entire electrical-power grid of the city during the early hours of the morning. For a few millionths of a second, thousands of megawatts were released into enormous lightning sparks. This produced a very small ball that bobbed about the vicinity of the spark, disappearing a few seconds later with a loud, sharp explosion. Had the ball been as large as some created under natural conditions, the entire laboratory would have been devastated.

There have been many hair-raising eyewitness reports of ball lightning. One was seen by terror-stricken airliner passengers. It is claimed to have actually formed *inside* the craft, where it silently floated along the gangway like a luminous phantom, eventually seeming to vanish right through the structure of the plane. A red lightning ball about sixty centimeters in diameter is claimed to have dug a trench a hundred yards long and more than a yard deep in soft soil near a stream and then torn away a further twenty-five yards of the bank. There have been reports of others shattering logs, cutting metal cables and causing other damage requiring surprisingly large amounts of power.

In April 1976 the scientific journal *Nature* described how a glowing purple-colored sphere with a halo of flame suddenly appeared over the gas cooker of a startled housewife in Staffordshire, England. Physicist Mark Stenhoff of the Royal Holloway College, Surrey, who investigated the incident, says that the object was about the size of a tennis ball and emitted a peculiar rattling sound and a singeing smell. It moved toward the woman, who instinctively brushed it away. Almost simultaneously the ball exploded. It burned a three-inch hole in her dress and left her hand red and swollen. It has been estimated that this fireball generated a temperature of two hundred degrees Centigrade. When the hole in the dress was laboratory-tested, the central portion had vanished and the gap was surrounded by an area in which the pattern had faded and the fabric was shriveled.

Other eyewitness accounts of ball lightning were quoted by Neil Charman, a lecturer at the University of Manchester Institute of Science and Technology, in the magazine *New Scientist* in February 1976. One man from Norwich described how a glowing orange ball about two inches across once drifted through his scullery window and hovered by the down wire of an electric light. It suddenly vanished "with a loud pop." Another man recalled how, many years ago, while sheltering from a violent thunderstorm, he watched a glowing sphere the size of a football enter an empty stationary tramcar, glide through it and emerge at the other end, where it exploded on the ground outside. The witness said he later found large holes blasted through the front and rear panels of the tram. Charman estimates that the ball must have dissipated at least a million joules during its destructive passage through the vehicle. One couple claimed that, following an intense lightning flash over Lake Geneva, they watched a ball "the colour and size of the Sun at noon" suddenly appear a few hundred feet in the air. "The edges were slightly fuzzy," they say. "It was about three miles away and drifting with a smooth, non-rotating motion like a balloon for about half a mile . . . and then vanished."

From hundreds of eyewitness reports Dr. Charman lists

the following properties of "typical" lightning balls, or "Kugelblitz" as they are often called. They are generally spherical or pear-shaped, with fuzzy edges, ranging in diameter from one to one hundred centimeters. They come in a wide range of colors, red, orange and yellow being the commonest. Their lifetime usually ranges from one second to over one minute. They can move either horizontally or vertically or just remain stationary. Many balls appear to spin. In some cases their sudden disappearance liberates heat. Some lightning balls show an affinity for metal objects and may move along conductors such as wires or metal fences. Others appear inside buildings, "passing through closed doors and windows with curious ease. Chimneys, fireplaces and ovens appear to be favoured haunts of these erotic objects." They can vanish silently or explosively, often leaving odors resembling ozone, burning sulphur or nitrogen oxide, and perhaps a faint mist or residue.

In 1936 a correspondent to the London *Daily Mail* told how a large "red hot" ball fell from the sky, struck a house, cut the telephone wire, burned a windowframe, then submerged itself in a four-gallon tub of water. The water boiled for some minutes after. The witness could find nothing in the tub when he searched it. The implied energy of the ball was about 10^9 joules.

Recently, after researchers scrutinized more than a hundred thousand records of lightning strokes, they found six photographs of strokes, ending above the ground, with balls at the end which appeared to be floating to the ground.

Though no one really knows what lightning balls are, these frivolous objects appear to exist either through some energy source within themselves or by some external energy source that somehow sustains them throughout their brief lives. In most recorded instances their explosive force corresponds roughly to that of between two and a half and twenty-five pounds of TNT.

One theory is that they might be a nuclear phenomenon. M. D. Atshuler and other researchers at the High Altitude

Observatory in Colorado suggested they could be caused by reactions involving the production of radioactive elements from the oxygen and nitrogen of the air. However, if this were so, that Staffordshire housewife who brushed one from her dress would probably have received a sufficient radioactive dosage to cause radiation sickness. Similarly, if Kugelblitz, or ball lightning, is caused by minute traces of antimatter, anyone in its vicinity would be expected to suffer burns from the release of gamma radiation. No evidence of this has been found, despite tests carried out by researchers from Oxford University.

Anthony Lawton suggests that this bewildering phenomenon might even be a micro black hole—possibly created by the tremendous energy of a particularly violent flash of lightning. Lightning itself is an enormous spark or pulse of electrical energy that generates a colossal magnetic field. This field envelopes the spark and compresses it like a vise. In most cases nothing unusual results from this pressure. But there could be instances where a super-intense lightning flash, lasting for a far longer period than normal, might crush the ionized oxygen and nitrogen atoms to a point where they pass the Schwarzschild Limit * and collapse into themselves to form minuscule black holes.

The smaller the black hole, the shorter its life span. One that could exist for only a few moments would be less than the diameter of a hydrogen atom. One such as this could produce an explosion similar to those reported during instances of ball lightning. A black hole of such incredibly small dimensions would account for ball lightning being seen to pass through the walls of buildings and the fuselages of aircraft. To such a tiny entity the pores in stone or metal would be like wide-open "windows."

The micro black hole would also explain other characteristics of ball lightning. By ionizing the surrounding

* In 1916 Karl Schwarzschild, a brilliant German mathematician, deduced that light could not escape from a body that was creating a sufficiently powerful field, but would bend back to its surface. This effect is known as the Schwarzschild Radius or Limit.

air with its tremendous gravitational field, it would create the familiar glow and possibly produce the crackling and hissing sounds as particles of matter were sucked in and then blasted out again at extreme velocities. Since the ionization radius would be much larger than the nucleus, the glowing mass would literally float on air in exactly the way ball lightning does.

So could some particularly massive form of ball lightning conceivably be associated with the Tunguska Event?

Lawton calculates that to cause such devastation would require a specimen of Kugelblitz 3,700 feet in diameter. In fact this is very close indeed to the estimated size of the Tunguska fireball. From eyewitness reports early researchers reckoned it to be about one kilometer (3,300 feet). This is so close that we cannot dismiss the possibility that the Siberian monster might have been a giant Kugelblitz.

But is it possible for ball lightning to attain such immense proportions? The most common ones are merely a few inches across. Lawton maintains it is possible, provided the ball lighting is composed in a particular way. He suggests that if the fireball is made up from dust particles bound tightly together by an electrical charge, it could achieve these huge dimensions. One such fireball a kilometer across entering Earth's atmosphere might disintegrate in a gigantic electrical flash that would produce the scorching radiant energy and searing shock wave associated with Tunguska and with nuclear explosions. (Full details of Lawton's calculations are in Appendix III.)

8

Nature's Deadly Roulette

> There is no reason whatever why a similar visita-
> tion should not fall at any moment upon a more
> populous region. Had this meteorite fallen in central
> Belgium there would have been no living creature
> left in the whole country. . . . Had it fallen on
> New York, Philadelphia (nearly a hundred miles
> away) might have escaped with only its windows
> shattered . . . ; but all life in the central area of the
> meteor's impact would have been blotted out in-
> stantaneously.

That, according to one of Leonid Kulik's lectures in
Moscow, is the horrifying significance of the thing that
struck central Siberia in 1908. The only reason it did not
perpetrate one of the most appalling massacres in human
history was simply the "luck of the draw." Like a flaming ball of fortune, the object just happened to slot itself
into the spinning roulette wheel of Earth at a spot completely devoid of humanity. Had its timing been slightly
different, its trajectory a few degrees either way, the
Cauldron of Hell could have been the center of London,
Moscow or New York. It is estimated, for example, that

had it arrived four hours and forty-seven minutes later it would have crushed the mighty city of St. Petersburg.

Tunguska therefore gives us little ground for complacency. We don't know what the monster was. We don't know if others of its awesome ancestors visited us in the past. We don't know if and when its successors will strike again. All we do know is that it was capable of more destruction than anything that man's most devious ingenuity has so far created.

How many of its kind have pounded into the vast oblivion of the sea? How many have blasted tidal waves of sand across the empty deserts? How could we know, after the camouflaging winds of centuries have long since repaired their imprints?

And let's face it, if it hadn't been for the natural moat of atmosphere that envelops Earth, the bombardment of extraterrestrial missiles would have smashed our tiny fortress to bits long ago. Even those objects we know, the giant meteorites, could have pulverized us time and again. Imagine, for instance, the gargantuan dimensions of the one that left its moonlike scar in Arizona. It exploded with the violence of many nuclear bombs, even though a vast proportion of its original bulk must have been gnawed away by the hungry heat of its fierce atmospheric encounter. It's pre-entry weight must have been many millions of tons.

The fact that the outer defenses of our planetary environment was unable to ward off the Tunguska invader is particularly disturbing. The sudden and unheralded return of another in today's uneasy world could have even more far-reaching repercussions. At a symposium in 1964, James R. Arnold of the University of California remarked, "Let us hope that the next such impact occurs during a period of relaxed international relations. For it would closely resemble the explosion of an H-bomb." It is not difficult to foresee the dreadful chain reaction if it happened to fall at a time when the major powers were squaring up to one another with their itchy fingers poised over the red alert buttons.

In fact, on August 10, 1972, our planet had one of the luckiest escapes in its recorded history. A blazing bluey-white ball of fire, looking like a giant welder's torch, seared over the western United States and Canada at more than nine miles a second, leaving a sinister smoke trail in its wake. It was a thirteen-foot meteor and weighed a thousand tons. It was first spotted over Utah. One and a half minutes later it was a thousand miles away over Alberta. Its lowest trajectory occurred as it swept over Salt Lake City. The kinetic energy of this object has been estimated at twenty kilotons. In others words, it had the lethal punch of the atom bomb that destroyed Hiroshima. Even so, by giant-meteor standards it was comparatively small, its mass being only about two percent of the Arizona meteorite.

The startling thing is that, after finding its way here across billions of miles of space, the meteor missed hitting us by a mere thirty-five—and it was virtually a miracle that prevented it from devastating a crowded American city. For, instead of hitting the Earth—as almost ninety-nine percent of meteorites would have done—it literally ricocheted off the atmosphere and flew on harmlessly into space again.

Fortunately the Aerospace Corporation of Los Angeles had a satellite-borne near-infrared radiometer in orbit over the United States at the time which identified the intruder as a meteor. But had it not been spotted and had it in fact landed, the immediate assumption might well have been that it constituted a nuclear attack. By the time its correct classification had been confirmed by the White House, retaliatory warheads might have been on their way east.

The news of this frightening near-miss was considered so unsettling that details of it were kept secret from the public for nearly two years. They were not released until February 1974, when a short scientific announcement by R. D. Rawcliffe and others from the Aerospace Corporation appeared in *Nature*.

A further disturbing comment in the same issue pointed

out that calculations have shown that meteors of this mass should cross Earth's orbit, and therefore collide with it, an average of once *every ten or twenty years.* And remember, only one in a hundred of them "bounce" clear like the one in 1972.

One would like to assume that during periods of East-West détente neither side would take instinctive nuclear action over the sudden detonation of one of these stray cosmic bombs on their territory. But suppose some delicate state of international "call my bluff" was simmering at the time. Suppose the American President was poised over the threat of another Cuban-missile-style incident, or the Kremlin was seething about another possible Vietnam. What then? Would there still be the time and the inclination for cool, calm reflection?

It seems unlikely that anything could prevent a major confrontation if the intruding object happened to fall on one of two countries, allied respectively to the East and the West, that were already engaged in active warfare. It is chilling to contemplate such a dramatic incident taking place on Arab or Israeli territory at the height of a Middle-East war. By the time the true cause was discovered it would most certainly be too late. Nuclear warheads could be crisscrossing the world like scheduled international airway flights.

One of Britain's near-misses from "NFO's"—natural flying objects—is thought to have occurred millions of years ago in the rugged Shetland Islands, creating what is known as St. Magnus Bay. The bay, which is the most prominent feature of the islands, is ten miles across and 540 feet deep in the center—unprecedented depth for coastal water in that region. Admiralty charts show that the ridge around it continues beneath the sea to form a complete and almost perfect circle. Scientists from the University of Liverpool suggest it was excavated by a two-million-ton meteorite more than a thousand feet in diameter that struck the Shetland Islands at a velocity of twenty-five miles a second. They believe it exploded, sending pulverized material high into the stratosphere in the form of a giant mushroom cloud. The rocks in which St. Magnus Bay is formed were once joined to the Canadian Shield, in which geologists have traced some thirty impact scars since 1950.

Ironically enough, the Tunguska Event itself might well have taken place here, even though the islands are nearly four thousand miles away from the Southern Swamp of central Siberia. For if the monster had arrived in the vicinity of Earth some hours later it could have splashed down just off the Shetlands. *An even more startling possibility, however, is that had its course changed very slightly in longitude and been depressed by ten degrees in latitude (a very small variance) it would have exploded at latitude 51.5 degrees at shortly after 7 A.M. Greenwich Mean Time. Its target—the center of London.*

Consider how slight this variance would have needed to be. The simple reason the monster exploded over the Stony Tunguska region was because that particular part of the world just happened to be in its line of flight at the time. The Earth makes one complete revolution every twenty-four hours; this corresponds to fifteen degrees of turn in one hour. Tunguska is situated at longitude 101 degrees east. Greenwich is at longitude zero degrees. To move those vital 101 degrees, the Earth takes six hours forty-four minutes. So had the monster arrived from outer space that much later, it would have struck at longitude zero.

It exploded over Siberia at seventeen minutes and eleven seconds after midnight GMT. Had it done so at seven hours, one minute and eleven seconds instead, it would have been slap on the Greenwich meridian. On the latitude corresponding to Tunguska, this would have placed it slightly north of the Shetlands.

But let's assume it had steepened its trajectory by a mere ten degrees (or ricocheted off the atmosphere as the meteor did over the United States in 1972—but in this case dropped back toward Earth). It would then have exploded at latitude 51.5 degrees north and longitude zero. Only one point on the globe corresponds to these two coordinates—London. These same coordinates fed into the automatic-pilot device of any long-range missile would steer it to no other target but Britain's capital city.*

Had the fireball followed the southwest-to-northeast course plotted by Kulik and others, it would have been first seen over Toulouse and Bordeaux. Had it taken the southeast–northwest course suggested by Krinov, it would have passed directly over Paris with a thundering roar like continuous gunfire. Some fifteen minutes later people in the French capital would have been terrified by a rumbling explosion and tremors shaking buildings all around them. As they gazed along the horizon they would have witnessed the billowing "tombstone" of smoke and dust towering in a fifty-mile mushroom cloud from the other side of the Channel. It would also have been visible from other major North European cities such as Copenhagen, Amsterdam, Oslo, Frankfurt and Berlin.

---

* Had the trajectory of the Tunguska object steepened by a further ten degrees—i.e., a total of twenty degrees—and had the object arrived eleven hours and twenty-seven minutes later than it did, it would have exploded over New York at about 11:45 A.M. GMT, or around 7 A.M. local time.

9

Latitude 51.5 . . .
Destination London

*Seven A.M., June 30, 1908. It is clammy and airless as
the stifled passengers on a local train steaming into Frankfurt fan themselves with rolled-up newspapers and shuffle
uneasily on the hard carriage seats. They are early starters
in what is obviously going to be a long, hot working day
in the dusty city.*

*The sun is up. There is no cloud. A few travelers
glance at their watches. One minute past seven.**

*And then they see it—streaking across the blue sky like
a bolt out of hell. An immense, searing ball of fire. The
train grinds to a halt and the passengers scamper out
onto the track. For a few moments they listen to distant
rumblings. Then nothing more. The air is still again.*

*Like the passengers at Filimonovo Junction on the
Trans-Siberian Railway, they are nearly four hundred
miles from the place the monster strikes—the center of
Edwardian London.*

*At seven o'clock on this sticky Tuesday morning, the
busiest city in Europe is already rustling itself awake.
In the stables of the elegant mansion houses coachmen
in shiny black top boots prepare high-stepping grays for*

---

* Greenwich Mean Time.

their lordships' first summons of the day, as a few ankle-hugging skirts already flap sidesaddle along Rotten Row. In the suburbs, the well-to-do middle class are still lingering over the family breakfast, ceremoniously laid out on heavy mahogany sideboards—mixed grills under silver-plated domes and precisely-timed boiled eggs nestling beneath hand-painted china covers. In less affluent Battersea, overworked, underpaid clerks have already left the tidy rows of lace-curtained terraces to the fussy indulgence of mobcapped wives who dust rooms littered like pawnshops with bric-a-brac and family heirlooms.

But in the city center itself the streets echo to the impatient clatter of racy little hansoms, gigs, dogcarts and lumbering drays. The walrus-moustached drivers of horse-drawn omnibuses—bright red and packed inside and out with bowler-hatted City workers—nuzzle their panting steeds past newfangled electric tramcars, cursing the arrogance of overtaking motorcars that honk pompously like flocks of frustrated geese. In prosperous Piccadilly, blue-aproned lackies are already polishing up the brass-door opulence of palatial offices and high-class restaurants in readiness for the clients whose business deals over coffee and brandy will influence the capitals of Europe. Breakfast is being served at the newly opened Ritz Hotel. The cleaners are dusting between the rows of red plush seats at Drury Lane, the Gaiety, Her Majesty's and the lively little Tivoli where Marie Lloyd and George Robey titillated packed houses the night before.

There's a sale starting at Harrods. The window dressers are busily arranging opera cloaks at fifty shillings, real ostrich boas at fifty-nine-and-six and "unrepeatable" ladies' chemises at seven-and-elevenpence. Billingsgate market has been alive as cockles for three hours, its cobblestone approaches crammed with fishcarts and cursing costermongers. So has the tangy-smelling Covent Garden market, where the porters balance leaning towers of round wicker baskets on their heads as they step gingerly through a swamp of flattened oranges and slimy cabbage leaves.

Big Ben stops at one minute and eleven seconds past seven. *A blinding split-second flash of hell incinerates the city. The whole of Great London is suddenly grilled and pulverized out of existence. It is literally torn from the map. The intense sunlike heat of the monster's breath of fire melts the very stone of the city's mighty edifices. Within seconds the innards of the metropolis are ruthlessly ripped apart and then sucked up into the fearsome vacuum that results. The vaporized debris of the most noble buildings in the world—the Houses of Parliament, St. Paul's, Westminster Abbey, Buckingham Palace and the rest—is whirlpooled high into the sky to hover in a bellowing mushroom-shaped cloud.*

Shock wave follows shock wave. Destruction within a twenty-five mile radius is total. Death is instant. Charred bodies in their thousands are frozen like waxwork figures in the final postures of life—rigid in the microsecond of blind, instinctive flight or joined with others in the last erotic contortions of embrace. The macabre bleaching effect of the blast has left the tortured silhouettes of others on the surfaces of walls still standing, like a grim mural record of the worst holocaust in the history of the human race.

The sickening scenery of death in the crowded streets is horrific. Countless thousands of dismembered corpses— limbs amputated by their roots like the trees now strewn about Hyde Park—smolder among an angry sea of ruins. The city is a crematorium where animate and inanimate life is welded together in the vast melting pot the monster has created. An omnibus, its passengers, and the horses that strained between the shafts a few moments ago are all fused into one shapeless hulk of indeterminate matter. There are no broad streets anymore. Just furrows in the devastation. The mansions of Mayfair and the slums of Whitechapel and Bethnal Green have merged into an indistinguishable jungle of molten rubble where those still bewilderedly alive can only stare blankly into the face of death.

Crystal Palace, that exotic showplace of Europe, vanishes in a hail of fragments. In an area around it, scores of people lie dead or dying, their bodies pierced by arrowlike splinters from nearly a million square feet of shattered glass. The rigid skeleton of this once glorious construction is composed of brittle cast iron. It doesn't bend to pressure like steel. It disintegrates into fearsome razor-sharp shrapnel that explodes in all directions, guillotining the living in its gory path. Many who momentarily survive the original flash of hell are immediately cut down by these lashing sabers that sweep indiscriminately over a three-mile radius.

The carnage from this murderous kind of fragmentation also occurs at London's main glass-domed railway terminals like Euston and Paddington. As with so many other Victorian buildings, their very components—stone, brick, cement, cast iron and glass—literally burst apart in the blast.

Thirty-six miles from London (the same distance as Vanavara from the Southern Swamp) is Henley-on-Thames. It is the first day of the world-renowned regatta. Scores of little craft, bobbing about in readiness for the first of the events that have attracted thousands of visitors, are suddenly swept from the river as the hot galelike blast wave sears through the town, burning and ripping away the roofs of picturesque little cottages and fragmenting the fine old stained-glass windows of its centuries-old churches. Even here the casualties are high, as a mammoth tidal wave also lashes up the river, claiming its own helpless victims.

All is silent now. And strangely still. There are no nightingales in Berkeley Square. No pigeons peck at the stub that's left of Nelson's shattered column. The River Thames boils. Giant tidal waves have lashed scores of vessels clean out of the water. Their twisted wreckage litters the enbankment, and the only things left to float beneath Tower Bridge are the corpses.

A charred copy of the day's Times *ripples away from the hand that clutched it. "The King and Queen arrived back at Buckingham Palace yesterday after a visit to Lord and Lady Pembroke at Salisbury," reads one item. Now Edward VII is dead. So is Prime Minister Asquith. So are most Cabinet ministers and members of the government. And so is the seat of an empire. . . .*

Though none of this nightmare really happened to London, it is only by such a reconstruction that one can remotely grasp the magnitude of the force that was unleashed at Tunguska. A few degrees' variance in course, a few hours' difference in time: that's all it would have required to make that lurid report authentic. So let us now consider the monstrous aftermath of such a disaster.

What we must try to visualize is a major capital city, the most influential in Europe, attempting to cope with a nuclear explosion one thousand five hundred times more furious than that at Hiroshima in 1945, at a time when it was unequipped to deal with even a traditional bomb attack. Even one of today's international capitals, well aware of the lurking specter of a nuclear war, would take decades to recover from a sudden deathblow of those awesome dimensions. But we are considering a city at the turn of the century, complacently assured of its impregnability so long as England ruled the waves and the guards went on changing at Buckingham Palace. A densely packed metropolis whose only answer to a thirty-megaton holocaust would have been a fire-fighting force that still relied on horses to trundle it around, and a primitive medical service that knew nothing of the appalling aftereffects of blast and radiation.

Ironically enough, the shape and size of the area of total devastation around Siberia's Southern Swamp are almost precisely those of Greater London. *This* Cauldron of Hell would have formed an elliptical ring of death, reaching areas as far from the city center as Willesden, Finchley, Walthamstowe, Woolwich and Kingston-upon-Thames. There would have been little anyone could do for those pathetic survivors left to scavenge a few more days of life from the ruination and the festering decay all around them; little anyone outside this gangrenous open graveyard could do as the gruesome symptoms of radiation sickness began to whittle down the white blood corpuscles and gnaw at vital internal organs.

Within days central London would have become a crude, lawless jungle, its crippled, its dying and its hopelessly insane cowering like wild animals among the carnage. For those able to ward off death it would have been a ruthless battle of survival. Food supplies would rapidly putrefy in the hot days and nights that followed. The strongest might loot and murder to exist a few more weeks. For many of the weakest the only relief would have been the already corpse-ridden waters of the Thames. There could have been few vestiges of human dignity left in this once proud and powerful city.

Some of those people on the fringes of the cauldron might have dragged themselves into safer areas outside. But even here they would find little consolation or help. The chaos and suffering would still be on a frightful scale. They might well have been forced back like lepers by neighboring communities determined to preserve for themselves what paltry scraps of survival remained.

It must be appreciated that a nuclear explosion, or superblast, is vastly different from that produced by conventional bombs such as the type used in World War II—bombs containing from five hundred to five thousand pounds of high explosive. These created straightforward shock waves that traveled *outward* at the speed of sound, lasting for a few seconds. With a smaller range of bombs, damage was mainly confined to windows and doors which were usually blown inward—though some objects might then have been sucked outward by a secondary shock wave. Flash burns from these conventional bombs were negligible. Casualties were mostly the victims of flying glass and other debris, bomb fragments, crumbling masonry from buildings receiving direct hits, or the outbreaks of fire that swept through the wreckage later.

Grim though these wartime attacks were, few people were actually killed by the blast itself.

But consider the appalling effects of a midair nuclear superblast taking place a few miles above a major city (say at around twenty thousand feet). Initially there would be a blinding flash, instantly burning any object it enveloped *due to radiation energy alone.* Any living thing within a few miles' radius of ground zero would be vaporized or incinerated beyond recognition. Most human beings outside that immediate zone but still within the radiation's deadly embrace would still die from dreadful burns. Ironically, the victims in these areas could be reckoned the lucky ones: death would be ugly, but mercifully quick. It would not be so for those in regions adjoining the fringes of total annihilation—in the nightmare zone of the "half dead." Here would be mile upon mile of dismembered humanity.

In the outer areas people *inside* buildings might survive the initial blast, provided their bodies were not exposed to radiation through windows and other apertures. But, so far, we have considered only the casualties resulting immediately the explosion occurs. The most brutal part of a superblast would be yet to come. This is the enormous shock wave it generates. Assuming the explosion took place at twenty thousand feet, this wave—traveling at the speed of sound, or roughly one thousand feet a second—would reach ground level in about twenty seconds. Meanwhile, during that sinister lull following the initial death flash, thousands of temporary survivors from inside buildings would instinctively rush headlong into the open. This is when the new scorching shock wave of white-hot compressed air would strike, claiming countless more victims.

Remarkably enough, though everything would be devastated or burned within a concentrated area of the blasts, one tiny region would be preserved—as happened at Tunguska and Hiroshima. This would be a spot directly below the explosion, where the shock-wave fronts would be canceled out by the reflected waves from the ground. Here would be a tiny protected haven virtually unscathed by it all.

Even the two types of blast so far described are not the end of a superblast's fury. There is still what is known as "negative pressure effect" (NPE)—a force that can be perhaps the most devastating of all in built-up districts surrounding the central area. What happens is that the blast wave now reverses its direction to form a partial vacuum. This literally bursts apart buildings that are already cracked and weakened. They all explode simultaneously in one almighty upheaval. Those earlier survivors would almost certainly be caught up in this final outburst as rubble and fragments showered down on the molten and flaming debris left by the first two killer blasts. This negative pressure is caused by the rising columns of hot air which spouts skyward to form the familiar billowing purple-brown mushroom cloud. The macabre constituents of this cloud would include not only the pulverized remains of buildings but also those of humans and animals.

So the carnage from a superblast is in three parts. First, the flash burning from radiant energy. Next the burns and blast from the *outward-moving* superheated shock wave. And finally the explosive effects of the *inward-moving* negative pressure wave. These same effects must have occurred at Tunguska. It would have been the final NPE wave that tore away the surface of the taiga and spread the resulting dust over a large part of the Earth's atmosphere, causing those brilliant sunsets referred to in Chapter 1.

It is not easy to estimate the total death toll for which the monster would have finally been responsible had it blown up over London in 1908. It would have been many days or even weeks before anyone could penetrate the central area of the city, and months before any meaningful estimate could be made of the staggering number of casualties. The true figure might never have been known. Somerset House, with its exclusive records of Britain's population census, would, of course, no longer exist. Only by years of painstaking research through the relatives of missing persons could even the roughest estimate be obtained.

If another Tunguska Event were to occur over present-day London, the total number of fatalities directly or indirectly caused by the explosion might be something like twenty million, which is even greater than the number of Russian men, women and children estimated to have perished in the Second World War. The toll of lives in and around Edwardian London might have been between six and eight million, if we consider the overall British population as being roughly a third of what it is today. In other words, the nation's death rate would have been a grim one in three.

In those days too it would have been a considerable time before information reached the rest of the country. Although, even then, the world's most sophisticated city had a rapidly developing private and commercial telephone service (it had been the National Telephone Company since 1906), all exchanges in and around London would have been destroyed. A further phenomenon would have disrupted communications. It is one now associated with all nuclear detonations and is called "electromagnetic pulse" effects (EMP). This takes the form of a huge electrical surge that is produced at the same instant as the initial flash of radiation. The power released is stupendous—about fifty thousand volts per yard of air, which is ionized in the immediate vicinity of the blast. It is also accompanied by blinding flashes of lightning. Most, if not all, of the telephone and telegraph instruments up to at least a hundred miles away would have been fused and put out of action.

The fact that communication cables in the early part of the century were aboveground would have meant that virtually every telephone would be destroyed by the intense wave of electricity passing through the cables, including those linking Britain with countries abroad. Marconi's recently established "wireless telegram" service too would have ceased to function in an ionosphere so drastically disturbed that any SOS messages would have been undetectable.

Even today this powerful EMP force (many of its side effects have "top secret" military classification) would destroy every transistor radio for miles. The ionospheric disturbance would make short-wave broadcasting impossible for several days. Only the defense communication networks, specifically designed to cope with EMP, would survive.

An example of EMP was actually recorded at the time of the Sikhote-Alin meteorite fall in 1947 (see Chapter 4), when the mechanic working on a telephone pole felt a severe electric shock from wires that were disconnected.

The only means of getting news out of a stricken London in 1908 would have been by word of mouth. Even when the rest of Britain heard the sickening facts, there would have been little immediate relief work that could be done. The only movement in and out of the disaster area would have been on foot or horseback. Hundreds of miles of railway track, curled like well-grilled bacon rind in the heat flash, would have been utterly unserviceable. No roads would have been passable. In fact, the great city of London may have been as desolate as the Siberian wilderness for years. It is unlikely the Edwardians could have accomplished anything like the kind of rapid restoration that took place in Hiroshima and Nagasaki after the Second World War. And remember, those atomic bombs were fifteen hundred times less powerful than the force we are considering.

Apart from the immediate massacre from the blast, the devastated sewage- and waste-disposal system of those days would have produced waves of epidemics unsurpassed since the Great Plague of 1665. The city would rapidly have been infested with multitudes of rats, flourishing unrestrained among the countless tons of rotting human and other remains. Sometime later grotesque mutations could have been expected to break out in these animals, caused by the gamma radiation of the explosion. They might then have completely overrun huge areas of southern England.

But the disastrous consequences of it all would not have been confined to London. A political and economic shock wave would have reverberated round the globe and could have changed the whole course of twentieth-century history.

Consider Britain's lofty place in world affairs at that period. The British Empire was at its peak. Edward VII —one of Europe's principal peacemakers—would have died, along with most important members of the royal family and the government, and those civil and military chiefs responsible for running not only their own country but many other parts of the Earth still under its flag.

In today's nuclear age, governments have emergency plans for any similar man-made catastrophe. The carefully monitored climate of international relationship would probably allow time for a nation to evacuate its principal citizens, its offices, wealth and documentation, to areas of safety before disaster struck. But in 1908 there would have been no such warning. Suddenly the central government would cease to exist. The gold that then constituted the foundation of Britain's economy would have melted into the rubble and dust of what had been the financial fortress of the Bank of England. The diamond market of Hatton Gardens, the priceless wealth of the art galleries, the museums and many other public and private strongholds of Britain's heritage would have been lost forever.

With the nation in a helpless state of chaos and confusion, the Army would no doubt have taken over the government and inflicted rigid martial law on the entire country. A new temporary capital city would have had to be established, perhaps at Manchester, Liverpool or some other region even farther north. But Britain would have ceased to exist as a world power. She could no longer have controlled the wealth and resources of her vast commonwealth, where apposing factions might well have seized this unexpected opportunity to stake their claims with violence and bloodshed. With Britain immobilized, it is almost certain that Germany would have hesitated no longer in her move to dominate Europe. It had always been King Edward who kept his ambitious nephew Kaiser Wilhelm II in check; with Edward's sudden death there is little doubt that World War I would have started years before it actually did.

All this, and so many other drastic reshuffles of events at that time, whose repercussions would even be influencing the world today, could have been the outcome of that one event on a morning in June 1908.

In fact all that London suffered were phenomenal thunderstorms now thought to be due to atmospheric disturbances caused by the Siberian blast. The following is part of a news article that appeared in *The Times* on July 6: "In London [on Saturday, July 4] the storms raged with great severity in most districts. Buildings were struck . . . torrential rain fell . . . the thunder was frequent and the lightning of the most vivid and alarming description." The report then tells of many buildings being struck by lightning, some destroyed or set alight; of flagstaffs and roof beams being split to pieces, coping stones crashing into the streets, railway tracks demolished and animals killed on the spot. One man had his clothes set alight and another was temporarily paralyzed. One house was said to have been "filled with blue flame." At a manor house near Newmarket, a rector and his family described how "a terrific shock appeared to shake the whole house, temporarily stunning them." The *Times* report goes on:

> The building seemed incandescent. Realising that something serious had happened, the rector rushed upstairs to the bedroom where the youngest child was sleeping. He found the passages full of sulphureous fumes. He then went to the assistance of the cook who was lying stunned in the passage. [She had been struck by falling masonry.] She was in a state of collapse and the other servants stated that they had been blown into the passage by some terrific force. . . . A huge gap showed that a large portion of one end of the rectory had been demolished. Masses of stone-

work were distributed about the ground outside over a considerable radius.

So it seems that even from nearly four thousand miles away, Britain did not entirely escape the wrath of the Siberian monster.

10

The Angry Earth

Though the Tunguska Event is unique in recorded history, we can compare its effects with a known energy force that comes not from the heavens but from the very bowels of "hell"—deep below the crust of the Earth. It is the latent ferocity of the volcano—ever poised to burst open and engulf all those who dare to live in its shadow of death. Man can never entirely trust these dormant brutes simmering and slumbering on the face of our crowded planet.

Six years before the Tunguska Event, such a force of nature went berserk on the sun-drenched little island of Martinique in the clear, blue tropical waters of the Caribbean. It was almost eight o'clock on the morning of May 8, 1902. Saint-Pierre, a bustling port with a picture-postcard maze of narrow twisting streets and quaint multicolored stone houses, was still rubbing the sleep from its eyes. Suddenly a series of ear-splitting explosions ripped through the still morning air. Six miles away a small volcanic mountain called Pelée tore itself apart. A gigantic blast furnace of superheated gases burst through the side of the mountain and seared across the terror-stricken city with the fury of a tornado. Three minutes later Saint-Pierre was a blazing shambles and all but two of its thirty thousand inhabitants were dead or dying.

The following is part of an eyewitness report by the assistant purser on board the *Roraima,* one of the ships steaming toward the port.

As we approached Saint-Pierre we could distinguish the rolling and leaping of red flames that belched from the mountain in huge volumes and gushed into the sky. Enormous clouds of black smoke hung over the volcano . . . There was a constant muffled roar. It was like the biggest oil refinery in the world burning up on the mountain top. There was a tremendous explosion about 7:45, soon after we got in. The mountain was blown to pieces. There was no warning. The side of the volcano was ripped out and there was hurled straight towards us a solid wall of flame. It sounded like a thousand cannon.

The wave of fire was on us and over us like a flash of lightning. It was like a hurricane of fire. I saw it strike the cable steamship *Grappler* broadside on and capsize her. From end to end she burst into flames and then sank. The fire rolled in mass straight down upon Saint-Pierre and the shipping. The town vanished before our eyes.

The air grew stifling hot and we were in the thick of it. Wherever the mass of fire struck the sea, the water boiled and sent up vast columns of steam. The sea was torn into huge whirlpools that careened toward the open sea. One of these horrible, hot whirlpools swung under the *Roraima* and pulled her down on her beam end with the suction. She careened away over to port, and then the fire hurricane from the volcano smashed her and over she went on the opposite side. The fire wave swept off the masts and smokestacks as if they were cut by a knife.

The ship's captain was the only person on deck not killed outright, but he fell overboard from the bridge. The purser goes on:

The blast of fire from the volcano lasted only a few minutes but it shrivelled and set fire to everything it touched. Thousands of casks of rum were stored in Saint-Pierre and these were exploded by the terrific heat. The burning rum ran in streams down every street and out into the sea. This blazing rum set fire to the *Roraima* several times. Before the volcano burst, the landings of Saint-Pierre were covered with people. After the explosion not one living soul was seen on the land.

Many of the fearsome blast effects reported at Saint-Pierre that day bore remarkable similarities to those that were to sweep Tunguska and Hiroshima. Although the phenomenal heat completely carbonized many objects in its path, others, like some of the trees in the Southern Swamp, were left completely unscorched. Glass was melted and heavy iron bars were distorted, yet quite nearby delicate chinaware remained unbroken. Among groups of people swept by instant death, some had been stripped naked while others remained fully clothed. A box of unignited matches was found right next to a man's cremated body.

Léon Compère-Léandre, a Negro, was one of the city's only two survivors. There is no explicable reason why he should have escaped. All those around him died. His account of the blast effects is remarkably comparable with that given to Leonid Kulik by the Evenki Semenov at Vanavara (see Chapter 2). He says:

On May 8, about eight o'clock in the morning, I was seated on the doorstep of my house, which was in the southeast part of the city. All of a sudden I felt a terrible wind blowing, the earth began to tremble, and the sky suddenly became dark. I turned to go into the house, made with great difficulty the three or four steps that separated me from my room, and felt my arms and legs burning, also my body. I dropped upon the table.

The rest of his account concerns the appalling deaths of the other members of the household.

> . . . At this moment four others sought refuge in my room, crying and writhing in pain, although their garments showed no sign of having been touched by the flames. At the end of ten seconds one of these, the young Delavaud girl, aged ten, fell dead. . . . Then I got up and went into another room, where I found the father Delavaud, still clothed and lying on the bed, dead. He was purple and inflated, but the clothing was intact. I went out and found in the court two corpses interlocked; they were the bodies of the two young men who had been with me in the room. . . .

Another Negro, Auguste Ciparis, who was in the prison dungeon, was the other survivor of the awful Saint-Pierre catastrophe, though he was badly burned.

The explosive eruption of Pelée completely devastated an area of ten square miles, compared to eight hundred square miles at Tunguska. And yet it claimed that staggering death toll of thirty thousand. The reason is that this particular type of outburst, known as a Petéan eruption, is the most deadly of all volcanic activity. Instead of spewing out the familiar lava flow, the vent of the volcano becomes blocked by solidified magma. There is no way out for the volatiles and molten materials inside. So they begin to create an enormous pocket of pressure below the vent. Eventually the towering fury can be held back no longer. It blasts through the weakest part of the mountain's geological structure. Mount Pelée suddenly gaped open at one side, and the hurricane of searing gases and particles blasted out at more than a hundred miles an hour across the built-up area of Saint-Pierre.

In June 1912, sensational volcanic eruptions ripped the earth apart in a wilderness now known as the Valley of Ten Thousand Smokes in the southern coast of the Alaskan Peninsula. Their booming anger could be heard nearly a thousand miles away. More than 30,000 million tons of rock and other debris were scattered over fifty square miles, reaching depths of up to seven hundred feet.

Fortunately the nearest inhabited settlements were fifty miles from the outbursts. Yet even here buildings were crushed by the ten-foot-high drifts of dust and ash that had risen in an enormous cloud, turning day into night. Rainfall, condensing into sulphurous vapors, burned holes in clothing and scarred the skin. Fish died in the rivers, and birds dropped dead from the sky.

It was four years after the eruptions when the first scientific expedition went to the valley. It was sponsored by the National Geographic Society and was led by Dr. Robert F. Griggs, who wrote: "The whole valley as far as the eye could reach was full of hundreds, no thousands—literally tens of thousands of smokes curling up from its fissured floor. It was as though all the steam engines of the world, assembled together, had popped their safety valves at once and were letting off steam in concert."

If such an event had happened near a packed town or city nothing could possibly have survived. Dr. Griggs, in one of his accounts, says that if it had taken place on Manhattan Island in New York City the explosions would have been heard in Chicago, the fumes would have swept over all states east of the Rocky Mountains and acid raindrops would have caused burns as far away as Toronto. "Ash would accumulate in Philadelphia a foot deep," he says.

> As for the horrors that would be enacted along the lower Hudson, no detailed picture may be drawn. There would be no occasion for rescue work, for there would be no survivors. The whole of Manhattan Island, and an equal area besides, would open in great yawning chasms, and fiery fountains of molten lava would issue from every crack. This, disrupted by the escaping gases, would be changed into red hot sand, which, consuming everything it touched,

would run like wildfire through the town. The flow of incandescent sand would effectually destroy all evidence of the former city. In its deepest parts, the near-molten sand would probably overtop the tallest skyscrapers, though the top of the Woolworth Tower might protrude if its steel supports could endure the fiery furnace surrounding them. It is doubtful, indeed, if there would be any considerable ruins left behind to mark the site of the great city.

There are some ways in which the devastation in the Valley of Ten Thousand Smokes can be compared with the Tunguska Cauldron of Hell, though obviously the source could not have been the same. Not a single life was lost. Trees for some miles around were flattened. And for months after the eruptions brilliant-colored sunrises and sunsets were reported over a wide area.

Twenty-five years before the Tunguska Event, a sleeping monster suddenly reawakened after two centuries of peaceful slumber and devoured an entire island. Between 5:30 and 11 A.M. on August 27, 1883, four monumental explosions tore apart Krakatoa, situated in the Sunda Strait between Java and Sumatra. Its thunderous rejuvenation set up towering fifty-foot tidal waves in the Indian Ocean that smashed through towns and villages along the coastlines. On that single day more than 36,000 people died as they groped blindly in pitch darkness.

Krakatoa's ear-shattering roars were heard more than two thousand miles away. The vibrations of its wrath cracked the walls of buildings 150 miles from the outburst. And, as happened following the Tunguska Event, people in many parts of the world gazed in wonder at phenomenally brilliant risings and settings of the sun.

There was no reason to expect Krakatoa to go insane the way it did. There had been no eruptions anywhere in the region since 1680. But in May 1883 the pretty little group of tropical islands began to feel the first shudderings of the reviving dragon in their midst. A few sightseers peered in casual curiosity over the rim of its gaping mouth—like Sunday-afternoon visitors at a zoo. They listened to it roar and stared into its deep hot throat as it belched out steam. But no one was particularly alarmed.

By the middle of August, however, the ferocious symptoms of its fitful awakening began to increase alarmingly. And on August 26 a series of terrifying detonations were heard all over Java. Pumice began to rain down on towns and villages from a sky ripped by fierce flashes of volcanic lightning. No one slept that night, and a strange shower of phosphorescent mud covered the decks, masts and rigging of ships up to fifty miles away from Krakatoa so that they resembled floating phantoms in the gloom.

On the sinister morning of the twenty-seventh, the island of Krakatoa began to disintegrate as four titanic blasts spat debris fifty miles into the air. Some 200,000 million cubic feet of rock vanished into the sea. Houses were shaken two hundred miles away. In Burma, nearly fifteen hundred miles from the eruption, people heard what sounded like heavy gunfire. Even in south Australia the explosions and vibrations kept people awake.

Back in the vicinity of the Sunda Strait vast numbers of people stumbled helplessly through the black pall of deadly ash that suddenly enveloped them, and mammoth tidal waves lashed angrily through their towns and villages, leaving a pitiful trail of human flotsam in their wake. Ships were washed two miles inland and stranded thirty feet above the water level. An enormous cloud of volcanic material rose fifty miles into the sky and began to spread over the world. By the end of the month it was billowing over the Atlantic. Five days later it was reported over the Hawaiian Islands. By September 9 it had completely encircled the globe. Like a rainbow after a storm, it picked out the spectrum of the sun and painted the sky in breathtaking masterpieces of color. In India the sun was seen as greenish blue. In Sydney, Australia, the ever-changing hue of the sky merged through yellow, purple, crimson, pink and brown. The sun appeared green, blue, silver and coppery from other widely separated regions. The atmospheric effects of the dust cloud were seen in Britain on its third circuit of the Earth.

Devastating though the Krakatoa eruption was, a force estimated to have been five times as powerful is thought to have wiped out an entire civilization in ancient times. And some scientists now believe this Mediterranean land mass that was sucked into the sea was the site of Plato's legendary island of Atlantis.

Nearly three and a half thousand years ago Crete and its string of little islands were inhabited by one of the richest, most highly intelligent and sensitive people on Earth, the Minoans. Their majestic cities contained palaces and luxurious, air-conditioned homes in an age when Britain was still uncivilized. They produced breathtaking works of art and were superb ocean navigators.

Without warning a five-thousand-foot volcanic mountain on the beautiful island of Thera unleashed its unprecedented savagery with the ruthlessness of several hundred hydrogen bombs. The center of Thera vanished with tens of thousands of people into a sea-filled trough thirteen hundred feet deep. The segments that remained are now known as the Santorini Islands. The tidal waves from this unimaginable holocaust are estimated to have loomed seven hundred feet. They radiated in all directions, surging onto the Greek mainland and even reaching the African coast. On Thera itself, a flourishing, densely populated metropolis with dazzling white marble palaces that symbolized the elegance of the Minoan race was bulldozed out of existence. Sixty miles to the north the inhabitants of Crete perished in the same kind of cataclysm that was to engulf those residents of Java and Sumatra in 1883. Great two-hundred-foot waves must have surged up the mountainsides, leaving hordes of corpses scattered like seaweed on a shore.

In the 1960s a team led by Professor Angelos Galanopoulos of the Athens Seismological Institute excavated part of the stricken Minoan city that had not been claimed by the sea. This was followed by further diggings by a party from American institutions headed by Professor Spyridon Marinatos of the University of Athens. Much of the city was still preserved—the tastefully frescoed walls of palatial relics, the rooms and passages of well-built houses, solid stone bases of once lofty columns, superbly fashioned pottery in colors as gay as they were the day they were painted, cooking utensils and precious goblets.

Galanopoulos was intrigued to discover remarkable comparisons between this shattered Aegean island and the legendary Atlantis described by the Greek philosopher Plato more than two thousand years ago. Now he and other researchers are convinced that the two islands are one and the same.

Galanopoulos even believes that the Minoan catastrophe could also have accounted for the ten Biblical plagues that swept Egypt, more than four hundred miles away. The estimated dates of Exodus coincide with the destruction of Thera. The description of "water turning to blood," speculates this Greek scientist, might have been the deposit of rose-colored pumice from the eruption. The "plague of frogs" could also have been caused by pumice driving them out of the water to die on shore. He offers explanations for the plagues of "lice, flies, boils, pestilence and locusts" as possible environmental aftermaths of the volcanic outburst. The "hail, thunder, fire and darkness" mentioned in the Bible were all the kind of phenomena widely recorded after the Krakatoa explosion and would also have taken place at Thera.

It is thought that the few scattered refugees from the doomed Minoan empire—possibly those who were absent on sea voyages at the time—fled to the Greek mainland to escape the deadly volcanic fallout. Here they could have passed on much of their culture to the Greeks and provided the basis for their science, mathematics and architectural techniques.

Those volcanoes considered utterly extinct have been known to produce the most catastrophic revivals. Inside their gigantic "combustion chambers" alarming pressures can slowly build up for centuries or even thousands of years. Generations may have farmed and grazed their herds over the rich, florid slopes. Then suddenly comes the monstrous ressurrection.

Vesuvius had been long "dead" when the solidified lava that had belched into the Bay of Naples in prehistoric times was confidently chosen as a site for the vigorous city of Pompeii. Yet in A.D. 79 the entire city was entombed in a matter of a few hours beneath a steaming ocean of cinders and lapilli. And there it remained hidden until researchers began its remarkable disinterment seventeen hundred years later.

It was like opening up an Egyptian pyramid. Inside were the "mummified" casts of many of the city's twenty thousand inhabitants, frozen in the last instant of life like stark tableaux depicting the agony of that fatal day. The terror, the contortion, the very expressions of fear and despair on their faces, had all been preserved—encased in the timeless vacuum of molds of volcanic ash that had been soaked and then dried and hardened in deathly embrace about its victims. As bodies had decomposed, their detailed impressions remained within the stony tombs. Some were huddled in doorways or petrified in the desperate actions of flight. Others were clasped in someone's arms or still clutching valued possessions. There was the form of a dog still frantically biting at its chain. Sixty gladiators were found dead in their barracks, two still manacled in their cell. A richly dressed woman was also found alongside them.

This incredible graveyard—still being excavated to this day—is a unique record of the life of a thriving commercial city exactly as it was nearly two thousand years ago. There are the colorful, airy suburban villas with their loggias and terraces and the luxurious houses of the wealthy merchants, enhanced with exquisite murals and mosaics. There are the theaters, the baths, the mighty Forum, the temples. Lining streets that still bear the deep wheel tracks of Roman chariots are the shop counters, some still bearing the eggs and loaves on sale that day; the roughly plastered inns; the stifling cubicles of the harlots. The walls still bear the revealing graffiti of everyday city life: snatches from favorite poets, salacious slogans and caricatures, obscenities and threats, the names of gladiatorial idols and the touching pledges of devotion to someone who may never have had time to be loved.

The only known eyewitness accounts of the great eruption of A.D. 79 come from Pliny the Younger, nephew of Caius Plinius Secundus (known as Pliny the Elder)—a prominent Roman public figure who was one of the victims. Though the accounts do not specifically refer to the destruction of Pompeii itself, they are the only firsthand reports of the eruption that have survived to this day. Here are extracts from Pliny the Younger's vivid descriptions:

> Buildings all around us tottered . . . The chariots, which we had ordered to be drawn out, were so agitated backward and forward, though upon the most level ground, that we could not keep them steady even by supporting them with large stones.
>
> The sea seemed to roll back upon itself and to be driven from its banks by the convulsive motion of the earth. . . . A black and dreadful cloud, broken with rapid zigzag flashes, revealed behind it variously shaped masses of flame . . . like sheet lightning but much larger. . . . Soon afterward the cloud began to descend and cover the sea. It had already concealed and surrounded the island of Capreae [Capri]. . . . Night came upon us; not such as we have when the sky is cloudy or when there is no moon, but that of a room which is shut up and all the lights are put out. You might hear the shrieks of women and the shouts of men; some calling for their children, others for their parents, others for their husbands, and seeking to recognize each other by the voices that replied; one lamenting his own fate, another that of his family; some wishing to die from the very fear of dying; some lifting their hands to the gods; but the greater part convinced that there were now no gods

at all, and that the final endless night of which we have heard had come upon the world. . . .

It now grew rather lighter, which we imagined to be rather the forerunner of an approaching burst of flame (as in truth it was) than the return of day. However, the fire fell at a distance from us; then again we were immersed in thick darkness, and a heavy shower of ashes rained upon us which we were obliged every now and then to stand up and shake off, otherwise we should have been crushed and buried in the heap. . . .

At last this dreadful darkness was dissipated by degrees like a cloud or smoke; the real day returned and even the sun shone out, though with a lurid light, like when an eclipse is coming. Every object that presented itself to our eyes seemed changed, being covered with deep ashes as if with snow.

The sprawling hunk of Mount Vesuvius still reminds us of its occasional virility. Its minor, bilious spasms may be more predictable to today's scientists. But, as we've said before, you can never entirely trust a volcano—any more than the many other aspects of nature's restless moods.

The summer of 1976 was notable for a succession of natural catastrophes and dramatic climate anomalies. Banner headlines in newspapers throughout the world were asking such questions as "What's Happening to Planet Earth?" In Britain the *Daily Mail* listed a "Roll Call of Disaster" for serious earthquakes alone in the first eight months of the year. It read as follows:

> FEBRUARY 4—Guatemala, 22,419 dead.
> FEBRUARY 19—Cuba, one dead, eight injured.
> APRIL 9—Ecuador, 18 dead.
> MAY 6—North East Italy, at least 850 dead, thousands injured.
> MAY 17—Uzbek, Russia, no official loss of life reported, but strength of earthquake indicates probably heavy damage and death.

JUNE 7—Acapulco, Mexico, heavy damage, numerous injuries.

JULY 28—Tangshan, Northern China, strongest earthquake in the world in past 12 years. Tangshan devastated, Peking and Tientsin extensively damaged.

AUGUST 17—Southern Philippines, more than 3,000 dead [later figures claimed 8,000] in tidal wave following earthquake south of Manila. 50,000 homeless, major tremor followed by at least ten aftershocks.

AUGUST 17—Chengtu, Western China, rocked by major earthquake. [Note: The population had been evacuated before the quake.]

AUGUST 19—Four killed and 50 injured at Denizli in Western Turkey.

"Planet Earth, it seems, is on the rampage," the *Mail* article began, and, after referring to earthquakes, volcanic eruptions and worldwide weather aberrations, it asked: "Is something going wrong on a devastating, planet-wide scale?"

The question has provoked many conflicting views from geologists, astronomers and other experts. Many scientists are now convinced that the answer lies 93 million miles away in space—in the swirling inferno of the outer layers of the sun. For a two-year study by scientists at America's leading atmospheric research center at Boulder, Colorado, has now confirmed what was once regarded as an old wives' tale—that the activity of sunspots can drastically affect the world's weather. (Sunspots are seen as dark blotches or spots on the face of the sun which produce intense magnetic fields.) The researchers have found that during periods of minimum sunspot activity—as in the mid-1970s—droughts and certain other climatic anomalies are most prevalent in the West. Periods when this activity is at a peak appear to be wetter and more conducive to good crop harvests.

But apparently that's not all that these great solar flares can do to our vulnerable planet. It now seems that they can have a direct effect on what are known as the "solar winds"—forces consisting of streams of high-energy particles that are emitted from the sun at almost the speed of light, and which can violently interact with the Earth's magnetic field, causing electrical storms and powerful movements of air. Incredible though it may seem, the varying force of these air movements is now thought to actually increase or slow down the rate at which the Earth spins on its axis. And even though the variations are minute, they are believed to be capable of disturbing the sections of the Earth's crust that are interlocked like the pieces of a jigsaw puzzle and make up the land masses of the world (see reference to the Continental Drift on page 45). Such movements can then produce earthquakes in regions straddling the separate sections.

One theory is that the position of the planets in our solar system can have a vital influence on sunspot activity and, consequently, play a part in triggering off earthquakes. Two established scientists have, in fact, made the startling prediction that in 1982, when all nine solar planets will be on the same side of the sun, their combined gravitational pull will so dramatically affect the rate of sunspot activity that the crowded city of Los Angeles, lying on the famous San Andreas Fault, will be subject "to the most massive earthquake experienced by a major center during this century."

In their controversial book *The Jupiter Effect,* Dr. John Gribbin of the science policy research unit at Sussex University, England, and Dr. Stephen Plagemann, an American astronomer working with NASA, claim that this rare planetary alignment in 1982 (it occurs every 179 years) will "trigger off regions of earthquake activity on Earth, and by that time the Californian San Andreas Fault system will be under considerable accumulated strain."

Another American scientist, Dr. James Whitcomb of the California Institute of Technology, says the disaster could take place as early as 1978. The last time the San Andreas Fault was significantly disturbed resulted in the famous one-minute San Francisco quake of April 18,

1906, that killed seven hundred people. A similar upheaval today would claim infinitely more victims.

Despite the conviction of most seismologists that a major catastrophe can occur somewhere in California at any time, it is still one of the fastest-growing states in America. Speculators and big-time commercial operators are, of course, reluctant to publicize the obvious danger. It's bad for business.

California's precarious situation exists because one part of the region is "riding" on a different piece of the earth's continental "jigsaw" from the rest of the United States. As the strain builds up, the danger of the two sections slipping or jerking apart increases.

Could the rare alignment of the planets in 1982 produce that fatal "jerk"? Drs. Gribbin and Plagemann believe it could, and will.

> Some astrologers [they write] mark the beginning of a new age by the occasion of the grand alignment—when Jupiter aligns with Mars and the Moon is in the Seventh House, the Age of Aquarius begins. The Age of Aquarius will be, we are told, a time of peace and love. But will it be ushered in by a major slip of the San Andreas Fault and a wave of earthquake activity around the globe, unprecedented since seismology became a true science? . . . Most likely it will be the Los Angeles section of the fault to move this time. Possibly it will be the San Francisco area which has a major quake. The prospect of both these sections of the fault moving at once hardly bears thinking about.

Seventeen major earthquakes have occurred since 1836 within about 150 miles of the famous Golden Gate Bridge. Of these, eight were within a fifty-mile radius from the entrance to the bay. "Every one of these eight earthquakes occurred within two years of one or other of the dates of sunspot maximum," the two scientists point out.

A tremendous amount of research is now taking place in desperate bids to successfully predict earthquakes and so save many thousands of lives every year. The Russians have found that the seismic velocity (the velocity of sound waves) through rocks can vary prior to a quake. In China, ordinary citizens are urged to watch for any telltale signs from such things as abnormal animal behavior—like uneasiness in dogs, and birds suddenly ceasing to sing. Colored posters tell them to look out for "chickens flying into trees, rats moving house and pigs behaving like hooligans." There has been total response from the Chinese people. Historically their country has the world's worst record for earthquake deaths. It is on the Pacific earthquake belt, one of the two major belts which together encircle the Earth and on which ninety-five percent of all major disturbances occur. Despite this nation's vigilance, there appeared to have been no accurate prediction of the appalling Tangshan earthquake, one hundred miles from Peking, on July 28, 1976. It has since been estimated that more than 650,000 people died, although, with typical Chinese inscrutability, no official figures were released to investigators from the West. Eyewitnesses, however, spoke of an entire hospital and a crowded train disappearing when the ground opened up beneath them. For weeks after the disaster millions of Peking residents camped out on the streets following warnings of further possible tremors.

Almost a year after the Tangshan disaster, Cinna and Larissa Lomnitz, a husband and wife research team from the National University of Mexico in Mexico City, spent two weeks doing seismological research in China where they met scientists from the Tangshan area who had taken part in rescue work. Though denied access to Tangshan itself, the couple gathered remarkable data from eyewitnesses of the holocaust, some of which was reported in *The New York Times* in June 1977:

Just before the first tremor last summer at 3:42 A.M., the sky over Tangshan lit up "like daylight," waking thousands who thought their room lights had been turned on. The multi-hued lights, mainly white and red, were seen up to 200 miles away.

Half a mile from the fault line, which ran north-northeast and south-southwest, one field of corn the size of an airport was knocked over in the same direction as though by some giant wind. Leaves on many nearby trees were burned to a crisp and growing vegetables were scorched on one side as if by a fireball.

The powerful subterranean movement wrenched the surface earth several feet apart in some places . . .

The earthquake shock itself came with terrific suddenness. One man described it as a "huge jolt from below" that threw many people up against the ceiling. In Tangshan, which lay directly over the epicenter, people clinging to trees or posts were swung around by the swaying earth. Thousands of structures collapsed simultaneously "as if they were made of cards," a survivor said.

Thousands of sinkholes, shaped like bomb craters, appeared throughout the Tangshan area. Trees were snapped off or uprooted. Railroad tracks became tangled wreckage. . . .

An artist's impression of the Viking Mars lander as it prepares for touchdown on the Martian surface after its 440 million mile journey from Earth. (The Associated Press Ltd)

The almost total devastation of the city of Hiroshima after the explosion of the first atom bomb on 6th August 1945. (The Associated Press Ltd)

An old larch with young suckers growing around it in the north-west peat bog area of the cauldron. (Pergamon Press Ltd)

Continuous forest devastation near the bank of the river Khushmo, to the south of the cauldron. May 1929. (Pergamon Press Ltd)

The Evenok Il'ya Potapovich Petrov (Lyuchetkan), eye-witness of the fall of the Tunguska meteorite, and one of the guides of Kulik's expeditions. (Pergamon Press Ltd)

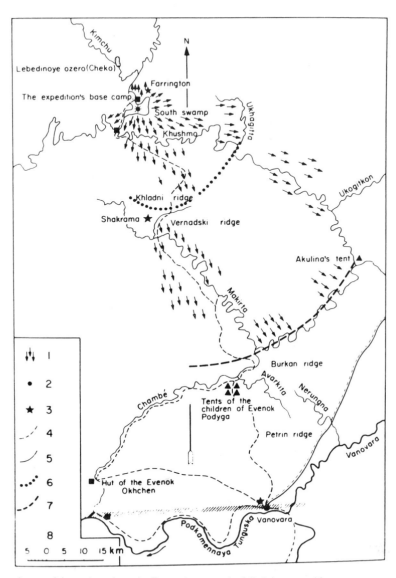

A map of the region where the Tunguska meteorite fell. 1 devastated forest;
2 place where meteor fell; 3 astro-radio survey points; 4 tracks; 5 road to the Strelka
trading station; 6 limit of scorched area; 7 limit of forest devastation; 8 limit of the
effect of the explosive wave.

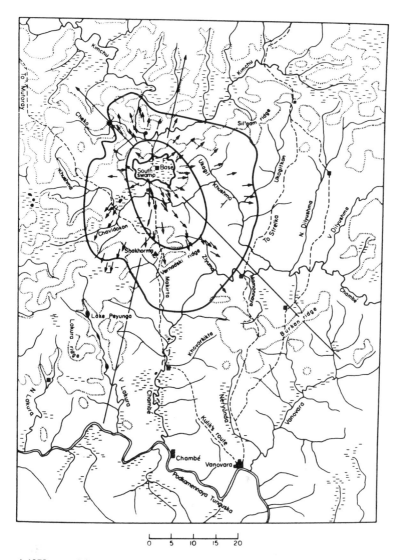

A 1958 map of the Tunguska region summarizing most of the data available at that time about the blast. The shape of the explosion wave and two different trajectories are shown.
(Pergamon Press Ltd)

A stony-iron meteorite from the Desert of Atacana in Chile. (Institute of Geological Sciences)

A polished and etched piece of the Canyon Diablo meteorite. (Institute of Geological Sciences)

A piece of the Canyon Diablo meteorite. (Institute of Geological Sciences)

A fragment of the Barwell meteorite ('stone'). (Institute of Geological Sciences)

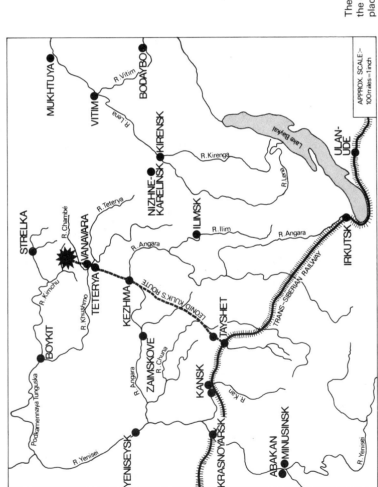

The Tunguska region in Siberia where the explosion (marked by a star) took place.

Part Three

11

The Alien Question

There is still one chilling question we must ask in searching for the source of the mysterious force that was unleashed over central Siberia.

Suppose the Tunguska monster wasn't a *natural* object at all. Suppose it was not something conceived and aborted from the vast womb of outer space, like a comet or a black hole; not some cosmic "gallstone" like a meteorite or a chunk of antimatter wrenched from the innards of a distant star system. *Suppose it was, in fact, artificial*—created by some form of intelligence.

We know that at that time there was no human agency remotely capable of producing a device of such multimegaton fury. And even if there had been, there is no plausible explanation of how or why it was unleashed over the desolation of central Siberia.

One must, therefore, ask the second startling question: If it *was* artificial, could it have been brought here by some form of interstellar space vehicle from a planet millions of miles beyond our own solar system—by an artifact that was pulverized in the split-second inferno that seared the taiga and perhaps committed the alien souls of a crew of highly intelligent beings to the Cauldron of Hell?

The idea is not a new one. Russian engineer and science writer Alexander N. Kazantsev has been fervently advocating this astounding concept since 1946. Soon after the atom bombs fell on Hiroshima and Nagasaki he was convinced that the Tunguska explosion was nuclear. He postulates that the atomic-powered engine of a fully manned alien craft, attempting to land on Earth, blew up over Siberia. Such a dramatic event, he insists, would have been entirely consistent with the many strange phenomena witnessed, and since investigated, at Tunguska.

Over the years, as each new expedition publishes its findings, Kazantsev slots them neatly and ingeniously into his spaceship theory. When globules of metallic iron from the impact zone were found to contain traces of nickel, cobalt, copper and germanium, he explained that the nickel and the cobalt originated from a splinter blasted from the outer shell of the spacecraft. The copper and the germanium, he declared, were from the electrical wiring, semiconductors and highly sophisticated instruments on board. Among his other arguments is that a spaceship exploding in midair would leave no crater. It would also explain the radial scorching and the undamaged trees at the very center of the blast. The fiery pillars and the billowing dust clouds too were consistent with the "mushroom" formations of Hiroshima and other nuclear explosions.

He describes the effects of a shattering spaceship as follows:

> The explosion wave rushed downward, and the trees directly below the point of the explosion remained standing, having lost only their crowns and branches. The wave burned the points of those breaks on the trees, and hit the permafrost, splitting it. Underground waters, responding to the tremendous pressure of the blow, gushed up as those fountains seen by natives after the explosion. But where the explosion

wave struck at an angle, trees were felled in a fanlike pattern.

At the moment of the explosion, temperatures rose to tens of millions of degrees. Elements, even those not involved in the explosion directly, were vaporized and, in part, carried into the upper strata of the atmosphere, where, continuing their radioactive disintegration, they caused that luminescent air. In part, these elements fell to the ground as precipitation, with radioactive effects.

Kazantsev even believes that this radiation may have caused the deaths of curious tribesmen who ventured too close to the Southern Swamp soon after the blast. At the time, a few superstitious Evenkis were spreading a rumor that Ogda, their legendary god of fire, was behind the whole thing and that some people had been burned and · killed. There has never been any actual evidence of such fatalities.

In 1959, Professor Felix Zigel, an aerodynamics expert at the Moscow Aviation Institute, agreed that the Tunguska explosion must have been nuclear. "At the present time," he wrote in the publication *Znanie-Sila,* "like it or not, A. N. Kazantsev's hypothesis is the only realistic one insofar as it explains the absence of a meteorite crater and the explosion of a cosmic body in the air." But, at that stage, he couldn't quite see eye to eye with Kazantsev's spaceship theory—especially as Kazantsev was suggesting that the spaceship came from one of the planets in our own solar system. However, as space research advanced dramatically in the sixties and more and more hardheaded scientists accepted the high probability of intelligent life existing in many parts of the galaxy, Zigel was more ready to believe in alien spaceships. In fact, in July 1975, now fifty-five and having helped train many of the Soviet cosmonauts, he stated publicly, "Based on hundreds of sightings reported by respected people and actual evidence taken to our laboratories, there is no doubt that extraterrestrial probes have been reconnoitering Russia."

Kazentsev interprets research by a Soviet aircraft designer, A. Y. Manotskov, as further evidence that the Tunguska invader was under intelligent control. Manotskov produced a detailed chart of the object's movements which he presented to Boris Laipunov, a Russian rocket expert and another spaceship advocate. They both agreed that as it neared the Earth the object slowed down to 0.7 kilometers per second, or about fifteen hundred miles an hour. As the average meteorite hits the ground at many times this velocity, could this indicate that the object was attempting to land?

Kazantsev also claims that, traveling at this slow speed, a meteorite capable of causing the Tunguska devastation would have had to be so vast ("more than one kilometer in diameter") that it would have covered most of the sky. Therefore, he insists, it must have been a relatively small body—the size of a spaceship.

Professor Zigel adds further intriguing speculation on the spaceship theory—that the Tunguska object made a deliberate "maneuver" and actually changed its course during descent. He and his scientific disciples base this conclusion on the conflicting flight paths calculated by experts from the varying descriptions of eyewitnesses. Voznesensky and Kulik had both fixed the trajectory as running from south-southwest to north-northeast. Later, however, Krinov and Florensky reckoned it was from southeast to northwest. Zigel concludes that the object was, in fact, seen to move in *both* these trajectories and had described an arc of some 375 miles in extent (in azimuth)—a performance that, he says, could have been achieved only by something under intelligent control.

Dr. Aleksei Zolotov, another scientist who visited Tunguska in 1960, attributes the oval shape of the blast area, shown in aerial photographs, to explosive material being encased in a shell. He reminds skeptics of the original eyewitness reports of "a body in the form of a pipe or cylinder" and "an elongated body narrowing toward the end." In 1975, Zolotov, head of a Soviet team still studying the Tunguska phenomena, embellished the theory by suggesting that the explosion was not an accident. It was, he suggests, deliberately detonated by a highly advanced civilization simply wanting to let us know of their existence; they chose the lonely Siberian site so that no human lives would be lost. A similar blast, he warns, "can be expected somewhere on Earth at any time." The doctor's startling diagnosis is based on the type of blast damage, and on such things as "evidence of the radioactive isotope cesium 137 in the ring structures of trees at the site."

In 1964, two Russian writers suggested that the Tunguska event was the dramatic result of a superconcentrated laser beam having been directed at Earth by an advanced intelligence occupying a planet in the star system of 61 Cygni—eleven light-years away.* Genrikh Altov and Valentina Zhuravleva, in a lengthy article published by the magazine *Zvezda* and summarized by the Soviet news agency Tass, say the laser beams were intended as signals and were sent here on three occasions—1882, 1894, and 1908. Evidence given for the first signal are the sightings of long green sausage-shaped "balloons" of light and large greenish clouds, particularly prominent over London and southeast England in 1882. The second laser signal, in 1894, produced similar phenomena, the two writers claim, phenomena which were reported at the time in the magazine *Nature*. The third beam, powerful enough to devastate Tunguska when it struck the ground, was apparently sent in reply to what the alien astrophysicists thought was a man-made signal from Earth. This "signal" was in fact the mighty volcanic eruption at Krakatoa in August 1883; the distance of eleven light-years between their planet and ours would have accounted for the delay of twenty-five years between signals. Altov and Zhuravleva suggest that material from the Krakatoa catastrophe, flung high into the air, reacted with

* A light-year is the distance that light travels in one year at a velocity of 186,272 miles per second. The distance of 61 Cygni from Earth is, therefore, 64.6 million million miles.

the upper ionosphere and could have emitted a strong radio beam and light signal far out into space.

The reason the two Russians selected 61 Cygni as the source of the laser beams is that one of its two stars is believed to be orbited by a very large planet about ten times the mass of Jupiter. In 1964 it was the only system known to have a planet of this size. Other, nearer stars have since been found to also have what appear to be planets. Barnard's Star, only six light-years away, is calculated to have three orbiting bodies as big as Jupiter; and the nearest system to Earth, Alpha Centaurus at 4.3 light-years, has a star identical to our own sun, extremely well suited to support an Earth-type world.

However, fascinating though Altov and Zhuravleva's theory may be, it is inconceivable that a laser beam could produce a midair explosion—certainly not one on the scale that flattened the Siberian taiga. If one actually struck our planet it would probably drill a hole in the earth, more like those caused by normal meteorites. But, instead of leaving the familiar untidy gap littered with shattered rock fragments, a laser-beam scar would be smooth where high temperatures had melted the rock. In any case you would get none of the pre-explosion "fireball" phenomena and streaks of brilliant light described by observers in 1908. The laser would pass through the atmosphere without trace until the solid surface of the Earth blocked its beam of energy and caused it to explode.

It would also have to be an unbelievably strong laser to cause such damage after traveling billions of miles, requiring an energy output many times higher than that of the sun itself. There is, of course, the remote possibility that an intense beam released from a spacecraft quite near to Earth could have created havoc. Though our own scientists are only on the fringe of laser technology, some American researchers say they will have the means of "wiping out an entire forest" by 1980. Already, under laboratory conditions, high-energy rays have burned through steel plate, and the canopy of a high-altitude aircraft is said to have been blasted away by a laser directed at it from the ground, though convincing evidence of the latter claim is not available at present. Vast sums of money are being allocated for laser development in both America and Russia in the race to perfect the first operational "death rays," which, among other uses, would instantly destroy guided-missile warheads. The 1976 U.S. budget for laser experiments was $170 million.

But if an alien society wished to communicate with us, why should they use destructive laser beams? That's no way to begin a lasting interstellar friendship. Mind you, should the intention of an alien race be to demonstrate its superiority with an initial show of strength, there are plenty of methods they could adopt. Apart from the use of their own sophisticated armory, one would expect they would be well able to harness the latent forces of our own geophysical environment in a number of devious ways.

Man himself is already learning to do this. In 1976 the Stockholm International Peace Research Institute (SIPRI) gave frightening details of possible warfare techniques for artificially modifying the weather and for triggering off earthquakes and other natural phenomena as a means of inflicting death and destruction. The previous year Leonid Brezhnev, secretary of the Soviet Communist Party, warned the world: "The level of contemporary science and technology is such that a serious danger arises that a weapon may be evolved even more terrible than nuclear arms." And the Russians have called on the United Nations to ban the "deliberate manipulation of natural processes of the earth, oceans and atmosphere."

The earliest of what are now termed "environmental and weather weapons" was used by the U.S. Air Force to "seed" rainclouds with silver iodide over the Ho Chi Minh Trail so as to stimulate torrential rainfall—though their available technique at that time was not very effective. They also created "Tunguska-type" destruction in the forests of Vietnam with herbicides and defoliants and by blasting huge tracts with "instant jungle clearance bombs." Russia claims that America is now experimenting with ways of inducing artificial earthquakes and volcanic eruptions by underground nuclear explosions, and of creating man-made floods, tornadoes and lightning. It is quite likely that a society of aliens whose technology is thousands of years ahead of ours could have perfected these and even far more appalling environmental weapons.

Two possible means of mass destruction now said to be occupying the minds of modern army strategists would seem to come straight from the realm of science fiction. One is to melt the polar icecap to create massive tidal waves. The other is to blast great gaps in the ozone layer of Earth. This would allow through the sun's deadly unfiltered ultraviolet rays, which could destroy vast numbers of people either through burning or through skin cancer. The main drawback to both these hair-raising techniques is that it would be difficult to confine their horrific effects to a particular "target area"; other parts of the world could be drastically affected, too. But, of course, a belligerent extraterrestrial intelligence would not be concerned about distinguishing between one nation and another. We'd all be the same to them.

One unsavory field of experimentation now causing concern to some UN delegates is the possible use of devices for producing *sounds* that can cause suffering and death. Tests with extreme low-frequency and infrasonic sound have been taking place in France for some years. Italian scientists are said to have killed laboratory animals by exposing them to high-frequency sounds; they found that each species was sensitive only to its own particular frequency.

There might be a number of ways in which an alien aggressor could effectively poison our atmosphere or even drain it of oxygen so that vast populations would suffocate. The North Vietnamese Foreign Ministry claimed that "oxygen-sucking asphyxiation bombs" were used by the South Vietnamese on the outskirts of Saigon (at that time a combat zone). During World War II many civilians in German cities suffocated when intense incendiary attacks drained the air of oxygen. Gases that attack the nervous system so that victims have no control of their body functions are also in the pipelines of the major military powers.

One particularly morbid form of mass subjugation that an advanced intelligence might possibly exploit would be to use the natural wave-guide between the ionosphere and Earth to propagate very low-frequency radiation in a way that would affect the electrical behavior of the human brain. "It might become hypothetically possible to impair the performance of a large group of people in selected areas over extended periods," says a SIPRI publication.

Immediately one considers the possibility of the Tunguska catastrophe being inspired by a "superrace," one must conclude that there must surely be any number of ways, far beyond the realms of our own modest technology, by which this could have been brought about.

Let us now discuss "evidence" for the spaceship theory as submitted by Kazantsev, Zigel and others.

First, the description of the Tunguska object as "a small body" does not fit that of any credible *fully manned* spacecraft propelled by normal nuclear reaction. To be capable of getting to our solar system from even the closest star (Alpha Centauri), such a craft, equipped with sufficient supplies and fuel for a two-way journey, would be massive—probably in the order of a hundred thousand tons. Remember that even traveling at the unlikely speed of light a star trek from Alpha Centauri would take 4.3 years plus time for acceleration and deceleration; traveling at a more feasible velocity, say a third that of light, would involve some sixteen years. For such a journey a full complement of maintenance and scientific personnel would be expected to total perhaps twenty aliens. The materials of which the vessel would have to be composed to withstand the rigors of such a trip through space could not be flimsy. One would expect it to be constructed of steel, titanium, cyramic or other such substances and deliberately designed to cope with immense stress, dramatically high temperatures and extreme radiation over long periods.

A ship of such rugged construction is unlikely to have been entirely vaporized by any internal explosion. The hefty shielding necessary to protect the crew from the radiation produced in the reaction chamber is unlikely to have disintegrated. Fragments of it should still have been found in an area which has now been combed more thoroughly over the years than any other region in the world. However, a really massive, sturdily built space vehicle such as this would account for it being seen and heard from places nearly four hundred miles away. And such an object *would* cause the brilliant tubelike streak described by eyewitnesses as it glowed in the atmosphere. The sonic boom as it plunged through would also explain noises like "artillery fire and thunderclaps."

This "tubelike" spectacle—described as being various colors by different witnesses—is consistent with the reentry effects of our own spacecraft. A. T. Lawton, in an interview with Rusty Schweickart, command pilot of the *Apollo 13* capsule involved in a dramatic NASA rescue operation, asked precisely what impression of the reentry an astronaut gets from inside the capsule. "It is fascinating," said Schweickart. "First faint violet tongues of flame appear around the edge of what seems to be a long blazing tunnel of light. At the end of it is just a jet-black patch of space. The flame becomes brighter, then glows first yellow, then brilliant green. It becomes so bright you cannot bear to look at it, and you are warned not to do so in case of eye damage. As entry proceeds, the color of the flame changes to red and finally dies out as you slow down to the speed of sound, and the parachutes burst open soon after."

However, it is not only spacecraft that produce these phenomena. A giant meteorite would do the same; in fact, a meteorite large enough to get through the atmosphere and retain sufficient bulk to cause serious damage—such as the one that pounded into Arizona—would create an entry tube several miles across and scores of miles long. The "cylinder shape" that a number of observers described could have been the entry tube of any rapidly moving body. Kazantsev and others suggest this was the actual shape of the spaceship. But the body could have been round or conical and still have produced a similar appearance.

It seems highly unlikely that an extraterrestrial crew would have been aboard a spacecraft traveling anywhere near Earth, as Kazantsev assumes. An unmanned "survey vessel" might have been dispatched, but it is probable that the mother ship would have been orbiting a considerable distance away. No sane commander, human or alien, would be likely to attempt to land the main ship on a strange, inhabited planet.

What *are* the chances of an out-of-control spaceship blowing itself to bits? For one driven by *conventional atomic power* they are extremely slim.

At the present time, the British Interplanetary Society is preparing "Project Daedalus," a remarkable blueprint for man's first nuclear-powered starship. This breathtaking concept is designed for travel to Barnard's Star system, six light-years away, at some fifteen percent of the speed of light, or 28,000 miles a second. Even at this staggering speed it is scheduled to take forty-seven years to reach its target. The journey would take a conventional chemical rocket around 135,000 years!

Daedalus would be gigantic—five times the length of a Saturn rocket, nearly twice the height of America's Empire State Building—with an overall weight of nearly seventy thousand tons. The propulsion would be by a nuclear pulse fusion system. In simple terms this would involve detonating thermonuclear pellets, or "bombs"— each the size of a tennis ball and each equivalent to a Hiroshima-type bomb—250 times every second. The colossal shock waves produced would thrust against a massive, flexible molybdenum mirror. This bowl-shaped reflector would then spring back, so transmitting the enormous thrust toward the main structure of the vehicle. The mirror itself, built to withstand white-hot temperatures of thirteen hundred degrees Centigrade, would, in fact, collect all the force of the explosion, projecting it in the right direction for the maximum propelling force. The nuclear fuel would be a mixture of deuterium and helium 3. A spaceship of this kind—and bear in mind it is designed for only a one-way, unmanned mission—would require forty thousand tons of fuel.*

True enough, contained in such a monster as this would be enough potential nuclear energy to blow up the whole of Siberia. In fact, if the complete second stage of a Daedalus "fly-by"-type spaceship hit the Earth the results would be diabolically more devastating than the Tunguska Event. Weighing some one thousand tons on its arrival and traveling at fifteen percent the speed of light, it would release energy 67,500 times greater in its collision! It would punch a giant gap in the Earth's crust, producing the worst series of earthquakes ever known and a sea of molten lava twenty-five miles across. The Tunguska Event, however, appears to have been the airburst of a fairly slow-moving object.

The chances of an uncontrolled blast on board Daedalus are virtually nil—just as they are with a conventional aircraft carrying nuclear bombs. There is no way in which these lethal pellets can be made to explode except by feeding them separately into the center of the ship's powerful detonating field and firing them with specialized gear. It is impossible for the rest of the fuel to be detonated spontaneously even if the spaceship crashed. One has to assume that any really advanced intelligence capable of bringing such a craft into our solar system would have ensured that it was equally accident-proof.

We cannot know, of course, the particular type of propulsion system any alien technologists might have developed. It could be very different from our own. But certainly, if the fuel content of a spaceship driven by

---

* Full details and specifications of the Daedalus Project are given in *C.E.T.I.: Communication with Extra-terrestrial Intelligence,* by the authors of this book.

conventional atomic means had exploded over Tunguska, hundreds of tons of uranium 235 would have saturated the area and been detectable in vast quantities. If the fuel had been plutonium the amount would have been somewhat less, but this material, one of the most poisonous elements known, would have had disastrous and long-term effects on human and animal life over a very large fallout region. Such appalling consequences could not have been overlooked.

We can calculate the amount of plutonium or uranium that would have been aboard an exploding nuclear-powered spacecraft to produce the equivalent thirty-megaton-bomb energy that blasted Tunguska. The atom bombs used against Japan to end World War II contained about six kilograms of plutonium or 18 kilograms of uranium 235. A thirty-megaton bomb would contain nine *tons* of plutonium or twenty-seven tons of uranium. However, to produce this energy release from a ship driven by reactors or piles, the quantities of plutonium and uranium would be about tenfold—i.e., 90 and 270 tons respectively. Such quantities of these materials would still be easily traceable in Siberia today. There is no record of such obvious evidence being found.

Here again an accident resulting in the kind of devastation at Tunguska is difficult to imagine. Every precaution would obviously have been taken in the construction of a spaceship needing to contain so much highly lethal material. It would certainly have been stored in many small, self-contained sections, each heavily shielded from the rest. Had an accidental explosion occurred in one of these compartments it is unlikely to have reached the rest of the system. And even if it did, this would not result in a catastrophic chain reaction. The vehicle would be more likely to melt under the intense temperatures produced, but a high proportion of the craft would fall to Earth intact. Large fragments, or even complete sections of it, should have been found at the impact site.

Kazantsev claims that traces of nickel, cobalt, copper and germanium found during expeditions were from a spaceship's instruments. But all these materials can be found in normal meteorites, nickel and cobalt being the most common ingredients. (See Appendix I.)

The reported presence of the radioactive isotope cesium 137 in the ring structures of trees cannot be accepted as evidence of a deliberately detonated nuclear explosion as Zolotov has suggested. It is true the isotope can be produced only by atomic fission. But this could have occured naturally in such an object as a comet head at the correct critical temperatures. (See Appendix II.)

Finally, we shall consider Professor Zigel's suggestion that the object actually "maneuvered" and changed its course during descent. There is no solid aeronautical evidence that it did. The theory is based merely on conflicting estimates of its trajectory from unqualified eyewitnesses in different areas.

But, even if Zigel is correct, even if it *did* make such maneuvers, this does not prove his contention that they could only have been carried out by an intelligent navigator. A comet could have done the same thing. Comets have frequently been observed to change course many times in distant flight. If a comet head composed of gases enclosed in ice had heated up on driving through the Earth's atmosphere, it could have burst on either side, causing a jet of gas forceful enough to push the object off course. The comet would then have described an arc for as long as the jet provided sufficient thrust. Comet Bennet was recently seen to break up as it approached the sun. Sections of it appeared to steer themselves about the sky due to this release of high-pressure pockets of gas acting like course-changing jets of a space vehicle. The gases released as a comet head dissolved could also produce a reverse thrust that would slow it down to the speed of sound.

However, despite all the uncertainties, until positive scientific evidence identifies the Tunguska monster beyond doubt, it would be unjustifiably dogmatic to entirely rule out the possibility of some form of spaceship having caused the devastation. The fact that such a bizarre event has never been recorded before or since, or is beyond our terrestrial comprehension, is insufficient grounds for dismissing the concept out of hand. If we do dismiss it, then we must also dismiss all the natural explanations put forward so far—for none capable of creating the same phenomena has ever been recorded, either. Even those who persist in the exploding-comet theory—admittedly the most likely natural cause—can turn up no records of other such happenings.

So what are the chances of such a visit from aliens? From all the evidence they seem to be extremely high, as we show in the next chapter.

12

The Living Universe

Since the installation of vast radio, radar and television networks throughout the world, the possibility of a visit from some extraterrestrial intelligence *must* have increased. Any such intelligence within a fifty-light-year radius of Earth should now be capable of eavesdropping on our transmission. And the further these penetrate the galaxy, the more new alien star systems come into range.

In November 1974 a three-minute message specifically intended for any receptive aliens was transmitted from the massive radiotelescope at Arecibo in Puerto Rico, modified to operate on wavelengths that advanced intelligences are thought likely to be using. The message, in an easily deciphered binary code, is a potted biography of the human race that also pinpoints our precise location in the galaxy. It has been beamed toward a cluster of 300,000 stars called Messier 13 in the constellation of Hercules. Though the signals, traveling at the speed of light, will take 24,000 years to reach this minigalaxy, there is still the fascinating possibility that one of countless stellar communities nearer Earth, but directly in the path of the transmission, might pick it up far sooner than we expect.

Carl Sagan of the Center of Radiophysics and Space Research at Cornell University told British television viewers in July 1976, "Though no one can be sure, it is beginning to look as if there are an enormous number of advanced technical civilizations out there in the universe. We have, for the first time in our history, the tools to find out." At Algonquin in Ontario, Canada, astronomers are using a 150-foot radio telescope for a systematic search of stars thought suitable for sustaining planets with intelligent life, in the hope of picking up signals. Russia too is well into the hunt for alien messages with her gigantic eighteen-hundred-foot reflector based in the northern Caucasus.

By the early 1980s NASA hopes to launch a 120-inch optical telescope into orbit around the Earth. Clear of atmospheric distortion, this $300-million watchdog of space, powered by direct energy from the sun, will probe galaxies of stars a hundred times fainter than those detectable by the largest ground-based equipment. And a specialized committee—on which Anthony Lawton, scientific editor of this book, represents Britain—was set up by the International Academy of Astronautics in 1974 to investigate all methods of communicating with extraterrestrial intelligences.

It is virtually certain that a vast number of these intelligences are already interchanging knowledge and ideas with each other over a farflung interstellar network, and probably have been doing so long before man arrived on Earth. Their technology could be so far ahead of ours by now that it would be utterly inconceivable to even our most accomplished scientists. The shock of actually discovering such a superior intellectual force anywhere within striking distance of Earth would shatter the complacency of the human race in a way never before experienced.

If any of those signals we are pumping into outer space does get a response, it must be treated with the utmost tolerance and respect. Alien minds could be quite impossible for us to comprehend. Their logic could be founded on quite different fundamentals than ours. Our philosophy and moral code may make no sense at all to them. The way they love, tolerate, or hate would have been influenced by emotional development tailored to cope with their own pattern of survival, just as their technology would have evolved according to the particular order of their enlightenment.

The manner in which they communicate may be based on principles vastly different from man's. Language, for instance, may be telepathic, visual or through some other medium rather than the spoken word. Their dominant senses would have developed in tune with their own environment—perhaps like the sonar senses of bats and other creatures on our own Earth. Man's concept of a friendly gesture might be interpreted by alien logic as a direct challenge. Our physical appearance might seem as aggressive to them as theirs to us. A really advanced race might have already had disastrous encounters with other aliens before contacting humans, which could make them instinctively distrustful of any uncustomary move we made. The last time they shook something's hand it may have clobbered them!

For these and many other reasons, man's first response to any extraterrestrial signal or approach would have to be guided by psychological rather than political instinct. For aliens to have been able to make contact in the first place would be a likely indication of their superiority. It would be suicidal to take chances with a superpower that might well be equipped to display its displeasure with sophisticated warheads lethal enough to blast us—and the entire solar system—out of existence.

The amount of caution necessary in the way we responded to any aliens would depend a great deal on how far away they were. Signals originating from a star merely a few light-years distant would demand the utmost diplomacy. The consequences of inciting the wrath of a superrace could be far more horrific than the force that struck Tunguska. George Wald, Harvard biologist and Nobel Laureate, is very uneasy about mankind making extraterrestrial contact. It would, he says "be the most highly classified and exploited military information in the history of the Earth." It is quite likely the general public would not be told.

Dr. Nikolai Kardashev, the eminent Russian astrophysicist at present searching the galaxy for alien signals at the Radio Institute at Gorky, has graded extraterrestrial life into three categories according to the level of their technology. He puts humans in almost the Grade I group—that is, among those societies advanced only sufficiently to utilize the resources of their own planet. His Grade II societies are those able to harness the total energy of a star, or sun. Finally there are the elite of the universe, the god-like races in Grade III who would command the unbelievable power resources of an entire galaxy—in control of millions of stars focused into a technology utterly beyond our wildest dreams.

The universe itself is a living thing, sensuous, hot-blooded and promiscuous. The intercourse of masses of plasma is continuous, its pregnancy perpetual—thrusting vibrant new suns into the galaxies so that they, in their turn, can nurture their own little families of planets. True, many of their offspring may be stillborn. Like the barren outposts of our own solar system they may be too far away to be breast-fed through the vital energy of their mother sun. But surely vast numbers of others will flourish into adolescence like Earth. Surely they too will have been safely nursed in a cotton-wool blanket of cloud, their steaming little bodies bathed by the cooling rain of centuries, their birthmarks topped up to become vast oceans. And surely nothing can then have stopped the molecules of organic life from taking shape.

In the Milky Way galaxy alone there are some 150 billion stars—suns greater or smaller, hotter or cooler than ours. Most are now believed to have planets bustling around them. And there are at least 100 billion other *galaxies,* each with as many stars or suns of their own. To realize the Earth's infinitesimal part in all this, consider that to make a one-way flight to one of the *nearest stars* by our fastest Saturn rocket would take an incredible sixty thousand years. For even *light* to reach Earth from the most distant stars in our galaxy (traveling at over 186,000 miles a second) takes 160,000 years. From the farthest known galaxies it takes ten *billion* years!

The total number of stars within range of our telescopes alone is estimated at 10^{28}. To appreciate the immensity of that figure, write down 10 and add twenty-eight noughts after it. And our observers could only yet have penetrated the fringe of the universe. It is only to be expected if those foregoing statistics are just too enormous to grasp. The human mind cannot do so. They are quite beyond its true comprehension.

Is it, therefore, in any way logical that all this cosmic matter, energy and fertility should have been created exclusively for the benefit of *one* race of moderately intelligent beings on *one* tiny planet? The purpose of life is to procreate. The universe is life itself. Its very elements are those of which every one of us is composed. So why should not those same elements have also combined into an endless array of other intelligent forms? It is arrogant and intolerant in the extreme to assume that the same biological evolution that culminated in intelligence on this Earth should not have found equal opportunity to flourish on other worlds. Even in the tenuous atmosphere of Mars the NASA Viking spacecraft detected a small percentage of nitrogen, one of the necessary elements in the creation of biological organisms.

For some years scientists have been able to simulate in their own laboratories the natural process of producing life by the simple interaction of certain basic chemicals. A gas mixture of methane and ammonia over water is subjected to radiated energy at levels believed to have existed on Earth in primeval times. The products resulting from these reactions are found to contain the very biochemicals required by our bodies for part of our metabolic process. Surely similar conditions and ingredients will stimulate the same life-producing reactions elsewhere in the universe.

It is now known that life forms can survive in the most outrageous conditions. Bacteria have been found flourishing in the hot springs of Yellowstone Park. Microbes discovered in antarctic core samples taken from thousands of feet below the surface have been thawed out and revived after half a million years in suspended animation. Certain fish in the polar regions are endowed with "antifreeze" elements in their blood. Microbes can live in acid, and bats in Mongolia exist on infinitesimal amounts of water. Why should extraterrestrials not flourish merely because conditions may not suit the cozy environment of delicate warm-blooded humans? Nature has the most miraculous ways of teaching all living things to adapt to the surroundings she creates. As Robert F. Freitag, the man responsible for planning NASA's future manned space flight program, says, "Today many scientists, if not the majority, agree that extraterrestrial life must surely exist, and possibly in enormous abundance. The question now is no longer so much one of 'if' as of 'where?' "

The one question no one can answer is in what form intelligent aliens have developed. The permutations are endless. Each species will be synchronized with its own particular environment, just as we are with ours. Their skeletal structures will have developed to cope with the territory and gravity of their own worlds. On a tiny planet the dominant life form might literally be "angels," light, willowy creatures adapted to fly over their low-gravity world, whereas great lumbering monstrosities might be the intelligent species that crawl slothlike across the face of a world whose powerful gravity would crush the human frame.

At the National Air and Space Museum in Washington are models of imaginary creatures from other planets. One, created to suit a high-gravity world, is a monstrous herbivore with eight legs, a large mouth in its chest, two eyes on a stalk and ears along the side of its body.

Its weight: thirty thousand pounds! Another, designed for a planet like Jupiter, is a great beast with a hydrogen-filled balloon on top of its head enabling it to float in the atmosphere.

Some of the really extreme genetic masterpieces of life in distant outer space would no doubt seem far more grotesque to us than even the most hideous creations from the fertile minds of the most outrageous writers of science fiction. The instinctive nausea and revulsion man could well experience when he meets his first aliens would be one of the most difficult psychological barriers to overcome. It would not be easy for him to disregard what, by his standards, is horrific far beyond his most tortured nightmares, and to accept such manifestations for their superb intelligence alone. It would not be easy to find himself exchanging information with, say, a loathsome, shapeless mass of protoplasm. *Somewhere* out there may be creatures similar in appearance to ourselves, but the odds against meeting them must be enormous.

Even man himself might eventually change as technology influences his environment. One group of scientists have predicted we may develop larger heads and smaller bodies. Increasingly efficient computers and labor-saving devices might result in smaller limbs, and our teeth might not be so conspicuous as we devour less and less solid food. Dr. James E. Harris, chairman of the Department of Orthodontics at the University of Michigan, has been studying the changing face of humans. He is convinced we'll eventually look different than we do at present. Our lower jaws, he says, will get smaller, eventually receding to a point where we have practically no chin.

But our aesthetic values will be adjusted along with our features. We will still consider ourselves to be as handsome as ever, just as the most repulsive alien we could ever imagine would be firmly convinced he was the gods' gift to the universe. The adage that beauty is in the eye of the beholder could not be more applicable than it is in any comparison of alien life forms.

For a moment try to imagine an alien's-eye view of late-twentieth-century humanity. We shall assume "he" is from a small, low-gravity planet and is a million years more technically advanced than ourselves. As we cannot possibly know what he would look like, we shall make him shapeless—just a delicate, pliable blob of near-transparent matter something like a huge jellyfish. On his world anything that did not have his own unblemished streamlined form would be unnatural and grotesque. Heads, torso, limbs would all be repulsively unfamiliar.

To him, humans would be strange, ungainly creatures. The face of a beautiful woman would be nothing more to him than a contorted oval-shaped chunk of protoplasm, hairy at the top, with two sockets and a flabby, red-rimmed hole that emitted peculiar grating sounds as it continuously opened and closed. Her elegant 36-21-36-inch figure would be an ugly, misshapen bulk with long, thin tentaclelike attachments which she occasionally wrapped around other malformed creatures of the opposite sex. Nothing in the general behavior of this curious species would be in any way logical to an alien mind that had formed in entirely different evolutionary channels. Human habits could appear disgusting and obscene—the way we eat and drink, the way we move, the way we make love.

Such an alien might belong to a highly organized community existing in a secure, utopian state of global togetherness that may have long since outlawed war, poverty and disease. Greed, suspicion and all social segregation through wealth, politics, religion and color might no longer fetter the unilateral progress of their planet. Human standards, ethics, values and loyalties could not possibly make sense to them. Nothing in the concept of their own clinical, finely adjusted logic could justify a single, dominant and reasonably intelligent species cowering in their incompatible warrens of philosophy and tradition—in most cases incapable of even understanding each other's languages, let alone each other's emotions and points of view.

The physical, moral and intellectual differences, between aliens and humans could be as vast as the chasms of space between their planets. In view of such unpredictability, in any sudden confrontation it would, therefore, be far better for friendly interstellar relationship if preliminary long-distance communication were first established so that each community could become mentally attuned and prepared for the kind of unfamiliar life form they could later expect to encounter.

13

The Intruders

In considering the exploding-spaceship theory in the search for the answer to the Tunguska Event, one must inevitably bring in the tantalizing controversy over unidentified flying objects that has raged all over the world for decades.

All those vivid descriptions of the Siberian fireball streaking across the sky can be found somewhere among the bulging files of UFO researchers. The "bluish-white light in the form of a cylinder" reported in the Irkutsk newspaper *Siber,* for instance, might be compared with the "hissing, cigar-shaped UFO . . . a brilliant blue-white glow surrounding it" claimed to have been spotted by a Police Sergeant Lester Shell in Tennessee in October 1973. However, there have been so many variations in the descriptions of "flying saucers" over the years that they could be made to account for just about everything that's ever been seen in the sky.

Author George St. George, who spent most of his childhood in Siberia in the early part of the century, makes a brief but fascinating reference to the Tunguska Event in his book *Siberia: The New Frontier.*

Some investigators seem to believe that whatever flashed across the taiga was intelligently directed because they feel only this explains the changing course of its flight. Was it then some sort of interplanetary vehicle in trouble, perhaps intentionally destroyed by its crew? Quite a few serious scientists seem to believe so. . . . Every serious UFO organisation throughout the world lists the Tunguska explosion in connection with possible interplanetary visitors who presumably have visited and studied our planet.

I remember this matter being discussed in our home in Chita in Transbaikalia, probably in 1914, by my father and his friend, a doctor who claimed to have visited the site of the Tunguska explosion a few months after it had occurred. The doctor had a detailed diagram showing the zig-zag course of the falling body over some 100 miles (where the tops of the trees were sheared off) before the actual explosion. He also said that some unusual glow was observed each night over the epicentre of the explosion for weeks after it had occurred, suggesting some sort of radiation.

My father, who was interested in the so-called "flying saucer" lore even then, was convinced that interplanetary visitors were using some parts of the taiga as their terrestrial base. He drew this conclusion from some ancient Evenki legends. [Of course the term "flying saucer" did not exist then, but the sightings of unidentified flying objects have been reported by the Siberian tribesmen, as well as the Mongols and the Chinese, for centuries.] Unfortunately all my father's voluminous notes on the subject were lost in China where he died in a Buddhist monastery in 1928.

The above extract is intriguing. If the author's reminiscences are correct, they would indicate that his father's friend possessed details of the Tunguska Event that could have been gathered only by visiting the site—such as the shearing of the trees. Yet, according to the author, he was discussing these things seven years before Kulik's first expedition.

Though nothing on the scale of Tunguska has ever been associated with the activity of UFO's, there have been many accounts of ground being scorched, of crops being incinerated and of odd indentations in the earth, sometimes attributed to spacecraft landing and taking off. In 1973 a farmer in Clarion, Ohio, found a forty-foot-wide area of his bean patch scorched, with holes a foot across at each corner. The previous year a similar discovery was made fourteen miles away by a farmer in Goldfield, where UFO watchers claimed an unusually large number of sightings in 1972. But such "evidence" as this would have been quite easy to set up by practical jokers—and there have been many of these responsible for "flying saucers."

But it isn't so easy to explain away reports of a "mini-Tunguska" from the village of Montauroux in the South of France in October 1972. This was the account published in the British *Sunday Express:*

Last Sunday (October 8) M. René Merle, a local peasant, went along to his woods for a little rough shooting, and was flabbergasted to find that 330 square yards of ground among the pines and white oaks had been "flattened."

Fragments of a section of dry stone wall had exploded in all directions, lacerating the bark of many trees. One tree stump, previously so firmly embedded that it could not be moved by man alone, had been uprooted and hurled several yards. Pine tree trunks, 18 inches thick, were coiled up as if by some gigantic centrifugal force, some twisted in a clockwise direction, others the reverse way.

A line of pine trees were sectioned as if by a blade in a cut rising progressively from 18 ins. to 6 ft. above ground level. On his dry wall, M. Merle detected traces of rubbing but no fragments of metal. The police, from the local gendarmes to Riviera headquarters, have confessed they are baffled. The wildest rumours are circulating this quiet corner of Provence.

Montauroux villagers are convinced that only a flying saucer could have wreaked such havoc to M. Merle's land which is more than 100 yards from the nearest forest track and bears no marks of a heavy military or other vehicle.

The newspaper report adds that a fireball was seen by several witnesses shortly before the discovery of the havoc. The fire brigade had been called out but had found no outbreak.

A chill early-morning mist shrouded the woods when Guy Turco, professor of mineralogy at the Faculty of Sciences at Nice University, arrived to study the phenomenon two days later. Souvenir hunters had already trodden and ransacked the area. He wanted to salvage whatever evidence might still have remained before a fresh horde of sightseers stormed in. After closely examining rocks and soil in the area he told the newspaper *Nice-Matin,* "I was searching for just a crumb. I didn't find it. I have nothing in my store of knowledge which allows me to explain the phenomena. I had come to verify the hypothesis of a meteorite. I found nothing which can confirm it."

Could the damage have been caused by a tornado, a whirlwind or, perhaps, lightning? Meteorological experts who later visited the site ruled out these possibilities. The trees had been twisted in different directions. A whirlwind always turns in the same direction, and it is unlikely such a force would have limited the damage to one circular area as had happened at Montauroux. "A whirlwind," explained one researcher "always has a definite trajectory. It does not begin and end in the same place. There would have been a trench, not a circle." Lightning was also ruled out by visiting scientists. They discovered no signs of burning. Though lightning can create devastating energy, they did not believe it could uproot tree stumps so deeply embedded in the ground or twist pine trees in that manner.

Among the investigators at Montauroux was a biologist from Cogolin, Alain Histarry. He claimed that directly over the stricken spot his compass "went mad." Instead of pointing magnetic north the needle swung firmly to due west. Professor Turco, however, could find no evidence of disturbance in the local magnetic field, and there were no anomalous amounts of radioactivity.

Within a week or so of the Montauroux incident little of the debris remained. Some fanatical collectors even took away the largest uprooted tree stump and later sold it to a local antique dealer!

Although these reports suggest nothing remotely as fierce as the Tunguska Event, the characteristics of the phenomena—the fireball, the flattened and uprooted trees, the angled slicing effect of the pines and the absence of fragments from any cosmic object—seem to come straight from the notes of Leonid Kulik more than thirty years before.

Since the 1950s Dr. Zigel of the Moscow Aviation Institute has been studying the many reported sightings of UFO's over the Soviet Union. One of the most extraordinary dates back to April 1961. Twenty-five witnesses claim to have seen a bluish-green oval-shaped object the size of an airliner traveling at tremendous speed over Lake Onega, northeast of Leningrad. A team led by scientist Fyodor Denidov investigated. Some eyewitnesses claimed that the object then flew so low that it seemed to scrape the ground, yet still continued on its course without slowing down. Major Anton Kopeikin, an army engineer who was also on the investigation team, reported that it left a trench in the earth fifty feet long and ten feet deep as well as two other smaller trenches.

Kopeikin told an American newspaper in July 1975, "The ice on the lake had broken up. The underside of the ice floes was bright green. Sample pieces of the ice, when melted, left a residue of magnesium, aluminium, calcium, barium and titanium. Also found were a strange piece of metal and tiny black grains of exact geometrical form consisting of iron, silicone, lithium, titanium and aluminium. The grains were resistant to acid and high temperatures."

Dr. Vladimir Sharanov, a Soviet geophysicist and member of the Leningrad Technological Institute, examined Kopeikin's findings. He is quoted as saying, "The black grains are unexplainable at the present time—but they are clearly of artificial origin. This object couldn't have been a meteor. As for the possibility that it was a plane, technical experts claim a plane couldn't withstand so heavy an impact against the frozen ground."

Among the case histories of UFO's in Zigel's file are reports of strange substances he calls "angel's hair" said to have been left floating in the air following some sightings. "It's usually like a glassy, cotton cobweb," he explains. "Witnesses say it has a bad odor and dissipates quickly. The secretary of the Committee for the Geophysical Year, N. V. Shebalin, has received several reports of angel's hair drifting through the atmosphere over the Voronezh region, south of Moscow. From the samples we've analyzed our scientists have not yet identified the substance with anything that exists on Earth—natural or man-made."

Similar material was claimed to have been found in the United States in 1974. Police departments in the Denver, Colorado, area received a flood of calls on the evening of October 16 that three UFO's (also tracked by radar at the International Airport) were hovering over a one-hundred-foot radio tower. At daybreak the following morning police claimed they found on the tower "strange, silvery cobweblike strands," which eventually blew away in the wind.

Zigel talks of another form of UFO "leftover" found by geophysicist Alexander Zayekin at Tambov, south of Moscow. He calls it "space tumbleweed" and claims that it contained "intertwining metal needles each about five to eight centimeters long and half a millimeter in diameter. The needles," he goes on, "were of an unknown brittle, gray-colored metal."

The discovery of such needles is intriguing. These could constitute tiny aerials whose length corresponds to a quarter wavelength of *an actual transmission frequency*. Strips of metal and needles of this type were used to confuse enemy radar installations in World War II. They act as signal reflectors and can sabotage attempts to identify aircraft. As they drift over a wide area they produce a confusion of echoes which makes it extremely difficult to trace the actual aircraft echo.

The fascinating thing about the needles Zigel says were found at Tambov is that those measuring five centimeters in length would correspond to a transmission wavelength of twenty centimeters. This is very close indeed to the "universal" wavelength of 21 cms., corresponding to a frequency of 1420 mhz. And this is one of the frequencies now being closely checked by astronomers all over the world in their search for possible signals from extraterrestrial intelligences. *It is a frequency scientists consider they could be using.* Needles measuring eight centimeters, also mentioned by Zigel, correspond to a wavelength of 32 cms., or a frequency of 938 mhz. This frequency does not fit any known *natural* molecular absorption frequency.

There are therefore two alternative explanations for the presence of these needles at Tambov. Either they are man-made and designed to correspond to frequencies used by the Russians in some of their military exercises or there is the fascinating possibility that they are extraterrestrial. Anthony Lawton explains: "If this were indeed the case, these needles might represent a riddle *and* the key to solving it, combined in one very neatly arranged package. The 5-cm. needles (or 5¼ cms., to be precise) could be recognizable as the clue to the 21-cm. 'universal' wavelength. The 8-cm. needles might then represent a frequency which could be used by an extraterrestrial group. They may wish to communicate on 32 cms., or 938 mhz., to avoid radio pollution at the more general frequency of 1420 mhz. In other words, though 938 mhz. is a totally unnatural or illogical frequency to use for interstellar communication, it has all the advantages of freedom from atmospheric and ionospheric interference and distortion." Using this hidden clue or key by deliberately mixing five- and eight-centimeter needles might possibly be an alien's way of testing the ingenuity of the human mind to ensure that it is sufficiently advanced to make further communication justifiable.

In most UFO sightings scientists are able to offer logical explanations. These include such things as unusual atmospheric phenomena, reflections from artificial lighting, heat mirages, high-powered xenon arc lamps on aircraft, Earth-launched satellites, weather balloons, and ordinary aircraft shimmering in the sunlight. Other causes can be hallucinations brought on by fatigue or mental stress, and strange effects on the human brain stimulated by electrical- or magnetic-field activity.

A twenty-three-year-old pilot who claimed to have been surrounded by UFO's on a flight to Mexico City in May 1975 was thought later to have been suffering from hypoxia, or lack of oxygen, a condition that might have stimulated hallucinations. Images and visions have actually been deliberately produced in the human brain by feeding slight electric currents into it. Certain people have also been found to be affected in similar ways by electrical, magnetic or radio fields such as those generated by lightning, overhead high-voltage cables or even intense sun-spot activity. The U.S. Air Force UFO investigation team can quote scores of other natural phenomena that can produce flying-saucer effects. They still admit, however, that about 2.5 percent of the thousand sightings a year that they scrutinized could *not* be explained by any of these phenomena.

However, if serious researchers in Siberia ever do find acceptable evidence that a spaceship blew up over Tunguska, it would certainly be the greatest-ever boost for UFO enthusiasts throughout the world.

14

The Incredible Answer?

Though all the explanations discussed so far in the "exploding spaceship" theory contain serious loopholes, there is one that would still account for every phenomenon recorded at Tunguska.

The hypothesis, however, is dependent on one vital factor: evidence for the presence of antimatter (see Appendix VI). If this *is* found, the possibility of it being *artificial* are overwhelming. For one cannot conceive how a piece of *natural* antimatter could survive long enough in our own galaxy to ever reach Earth. Only with some sophisticated form of protective shielding could such material be transported through the barrier of normal matter that makes up the environment of our solar system. And the most functional way in which this could take place would be in the form of fuel for a spacecraft powered by that devastating mixture of matter and antimatter.

In July 1976, a special NASA study into the prospects for future interstellar-space flight came to the following conclusion:

> There are those who speculate on the possibility of creating and containing antimatter. This would indeed represent the most efficient energy storage

system of which we can conceive, permitting the total conversion of mass to energy. If such systems could actually be operated as rocket engines, the flight time to a star might be cut down to a matter of decades.

However, we are not likely to see such engines before the year 2000. Although it is possible to create antimatter in high-energy accelerators, no-one as yet has any idea of how to store it in such a way that it could be made available as an energy source for a rocket engine. What is required is not simply further development along an established line of research, but a completely new breakthrough or invention. If this new invention—the storing of antimatter—were to come along in the next decade, then it might be technically feasible to build an antimatter-reaction rocket motor by the turn of the century.

A really advanced alien technology could well have acquired the delicately balanced technique of combining matter and antimatter in such a way that they could harness the awesome energy this would release. Basically there would have to be some form of reaction chamber where these two belligerent forces could be safely allowed to annihilate each other without blasting their surroundings out of existence, too.

The only kind of "reaction chamber" capable of containing such a highly lethal combination would be some powerful magnetic-force field which would also focus the released gamma radiation in the correct way for maximum thrust. The efficiency of such a propulsion system would be quite staggering and could drive a spaceship at enormous velocities exceeding ninety percent the speed of light. The antimatter fuel itself could not, of course, be allowed to make contact with any type of conventional container constructed of ordinary matter. So, here again, a magnetic-field system would have to be devised within which the antimatter could be safely isolated or suspended until it was required to be reacted explosively with the fuel of normal matter inside the reaction chamber.

The fearsome energy output from combining merely one pound of matter and one pound of antimatter would be equal to that produced by burning nine million tons of coal—enough to keep an average power station going for between five and eight years.

If the Tunguska Event *was* caused by a spaceship, the matter–antimatter drive model would seem the most feasible. If somehow the magnetic field isolating the antimatter fuel collapsed, this fuel would then make that fatal, uncontrolled contact with its ordinary matter surroundings. At the same time the power supplies controlling the other force fields would be cut off. The entire structure of the ship would be instantly vaporized without trace in a catastrophic explosion that could certainly have produced the type of devastation found around Tunguska's Southern Swamp region. In fact, if such a spacecraft was intending to return to its own planet, the quantity of fuel remaining on board would have been lethal enough to gouge out the entire Asian continent. It would require only a relatively small amount of antimatter to create the thirty-megaton blast over Tunguska.

This would indicate that the spaceship was never scheduled to return from its mission. Could it therefore have been manned by a small kamikaze group of scientists and technicians prepared to sacrifice themselves to the cause of interstellar research? If the explosion was accidental, were they intending to land when the force field suddenly broke down? Had they specifically chosen the uninhabited region of Siberia as a favorable location to set up base before making their first cautious contact with the human race?

A more likely alternative would be that the visitor was an *un*manned vehicle. In this case its computers could have been programmed to move it into a safe survey orbit round the Earth, but not to enter the atmosphere. Something might have gone irrevocably wrong during the complex maneuvers. The quantity of antimatter fuel retained to power an unmanned observation satellite of this kind would probably have been about the right amount to cause that thirty-megaton explosion. Another possibility is that the satellite was damaged in collision with some particularly formidable hunk of solar space debris, or that a vital system on board was drastically affected by a natural phenomenon closer to Earth. Whatever it was, once that force field containing the antimatter fuel broke up, nothing could have saved it from utter destruction.

The odds *against* a visiting spaceship from another star system containing living aliens are overwhelming. Why should it have been manned, when highly sophisticated, computerized satellites can do the job more efficiently? NASA did not send astronauts to seek for life on Mars in 1976. It sent those automated Viking orbiters and landers. The Russians didn't risk their highly trained space crews when they too attempted to investigate the mysterious Red Planet—which was rather fortunate, because their *Mars 3* lander was instantly knocked "senseless" by a fearsome two-hundred-mile-an-hour gale that suddenly lashed it across the rugged surface, putting its delicate instruments out of action. No space authority would commit human beings to the searing hellhole of Venus. But unmanned probes go there. In fact, two sent back dramatic pictures of the scalding planet while they were being slowly barbecued by its lethal atmosphere.

If *we* can investigate the solar system with these inquisitive little "private eyes," there is every reason to believe that advanced extraterrestrial intelligences are well capable of sending them far vaster distances. By the next century our own space program should be including these interstellar "messengers" that will feed back to mankind an unimaginable wealth of knowledge. From their lonely orbits around wonderful new worlds they will seek out vital clues to intelligent life, perhaps even intercepting alien radio signals. They will be telling us, too, which of those worlds might one day be suitable for the overspill settlements of the human race. Starship Daedalus is designed to be equipped with automated probes to investigate the planetary system of Barnard's Star. These would collect masses of data and send them to their mother ship for retransmission to Earth. But there would be no question of risking the lives of a living crew on these initial surveys of hostile planets.

Of course, had an extraterrestrial intelligence *already* sent probes to Earth and been now entirely satisfied that their own kind could survive here, then a "manned" expedition might be justified. There is the remote possibility that these earlier surveys found that human life was so primitive that the superbeings from their own planet could colonize it without serious risk. But the ratio of manned to unmanned space vehicles likely to be operating in the galaxy makes such a situation highly improbable. We might roughly calculate the odds from the breakdown of American and Russian missions to date. Of these (excluding military satellite launches) only one to two percent have so far been manned. On this basis, the odds against *interstellar* missions being manned would be much higher —possibly ten thousand to one. The enormous expense, the risk, the time element and the unwieldy size of a fully crewed starship all make unmanned exploration the obvious choice. In the early part of this century, a really advanced race might have been turning out assembly-line probes infinitely more efficient than the NASA Vikings at the same time the Americans were producing Model T Fords. And, despite their sophistication, just as cheaply.

In July 1975, an interstellar-exploration project that was submitted to the Committee on Science and Technology of the United States House of Representatives, and that now forms the basis for part of NASA's space program for the rest of this century, strongly advocated the use of automated probes to investigate nearby star systems. Among the findings, compiled by senior scientist Robert L. Forward of the Hughes Research Laboratories of California, are papers on alien-planet detection by Anthony Lawton. In his introduction to the project, Dr. Forward writes: "The initial goal is to send sophisticated autonomous probes to the nearest stellar systems before the century is out. The interstellar probes would use advanced computer micro-miniaturization techniques to achieve unrivaled orbital surveillance capability, combined with almost human intelligence, in a few kilograms of payload."

So could the Tunguska object, in fact, have been the counterpart of those Viking spacecraft that NASA sent to Mars—its mission, to seek out life *here* on Earth? With matter–antimatter propulsion power it could have streaked across a vast area of the galaxy at an incredible velocity. Once its highly developed computers had fed back their information to its own star system, the satellite would have been disposable. It might even have been programmed to destroy itself by deliberately collapsing the force field containing the last remnants of its anti-matter fuel. Alternatively some unforeseen system breakdown could have occurred that was beyond the ability of the computer "brain" to correct.

It might have ventured too close to Earth. Perhaps its auto-pilot system misjudged the height of the atmosphere and caused it to make entry. Here such a probe could not function. Its multitude of antennae, power-dissipating fins and other accessories sprouting from the main structure would have been designed specifically for the frictionless tranquillity of outer space. Once in the dense, restrictive atmosphere of Earth this vital equipment could have been torn away as the probe glowed like a giant multicolored fireball. It would be completely out of control. As in the reentry of man-made satellites, this searing atmospheric friction would consume the probe—its framework, motors, fuel tanks and everything else. In this situation a probe powered by matter–antimatter fuel *could* explode spontaneously. If the structure disintegrated in this way the antimatter fuel would be exposed and then be free to combine with ordinary matter.

One might even speculate on the type of fuel such a probe would carry. For various reasons it would probably be a liquid, a good conductor of electricity and preferably very dense so that it could be contained in a conveniently small area. Mercury would fit these requirements very well. One kilogram of antimercury could be contained in a tank about the size of a small thermos flask. This mini-tank would have an intense multiphase AC magnetic field surrounding it, thus suspending the fuel in a swirling mass in the center of the tank, safely isolated from the inner walls of the casing. Secondary fields at either end would hold the antimercury in a kind of "magnetic bottle." Adjusting the strength of the field force at one end would induce the fuel to enter a reaction chamber where it could be mixed with ordinary matter (again, probably mercury). If, in an accident of some kind, the magnetic field to the antimercury was cut off, the fuel would splatter the container walls in a flat "puddle." This would be more than adequate to trigger off an annihilation explosion.

Actually one kilogram (about two and a half pounds) of mercury–antimercury would produce an explosion far greater than that of a thirty-megaton bomb. So if a craft of this kind did destroy itself over Siberia the fuel content might have been less. This, however, is quite feasible, as it would require only a very small quantity to slow down a moderate-sized unmanned vehicle and to slot it into an orbit round Earth. The bulk of the fuel, of course, would have already been used up on the vast journey from its home planet.

Since matter–antimatter annihilation is the most powerful and efficient method of obtaining energy, it is highly likely that a technologically advanced race would be using it. As stated previously, in view of the insurmountable hazards of antimatter reaching our solar system in a natural and unprotected form, any traces of its original existence found at Tunguska by current or future investigations would be powerful evidence of the alien-artifact theory.

A satellite well equipped to transmit detailed information over vast distances to its own world may have been programmed to communicate with Earth had man's radio techniques been sufficiently advanced. But, of course, in 1908 they were not. Any attempt to create such a link-up would have been pointless. Realization of this might have been sufficient reason for the computer to release the order to self-destruct.

Until now advocates of the exploding-spaceship theory have assumed the vehicle to have been a massive one, driven by conventional nuclear power; a recent book *The Fire Came By,* written by Thomas Atkins and John Baxter, follows the popular reasoning that it must have had "a mass of thousands of tons" propelled by "nuclear fire." A small matter–antimatter–fueled probe such as we have described is a far more functional vehicle for aliens to use in any investigation of our solar system.

A number of scientists believe it quite feasible that alien probes such as this are exploring planets in many parts of the universe and could, at some stage, have orbited Earth. Certainly if such an object arrived here *now* our astrophysicists would be well able to set up a system of communication. A tremendous amount of research has been done in the last decade to cope with any kind of extraterrestrial signals that may suddenly arrive here.

One of the foremost advocates of the alien-space-probe theory is Professor Ronald Bracewell of Stanford University, who is one of the world's leading radio astronomers. In an article published in 1975 by *Mercury,* the official journal of the Astronomical Society of the Pacific (adapted from his book *The Galactic Club*), Professor Bracewell writes:

> It is clearly possible in principle to pack an enormous amount of information into a modest interstellar probe and to send it into the vicinity of a likely star. When the probe arrives, it fires a rocket and goes into circular orbit about the star in the middle of the habitable zone and in the star's equatorial plane—the plane where planets will be situated . . . the probe then begins a program of investigation, supplementing its internal power source with stellar power in a way similar to, or better than, the scheme that has been so successful on Skylab and other solar-powered spacecraft.
>
> It is quite possible that a probe arrived in our

solar system when our radio transmissions had not yet begun. It would therefore be desirable for the probe to continue listening. We are sensitive to this point, because the earth could have been the target of a messenger probe as recently as a few decades ago, but even then we did not have commercial broadcast transmitters in operation.

We would have a great deal to learn from such a probe. In fact, it could profoundly change our entire philosophy, uproot our religions and certainly destroy our chauvinism. But, as Professor Bracewell goes on:

> Knowledge of our language will enable the probe to tell us many fascinating things: the physics and chemistry of the next hundred years, wonders of astrophysics yet unknown to Man, beautiful mathematics. After a while it may supply us with astounding breakthroughs in biology and medicine. But first we will have to tell it a lot about our biological makeup. Perhaps it will write poetry or discuss philosophy. Perhaps the messenger knows how the universe started, whether it will end, and what will happen then. Maybe the probe knows what it all means.

If an orbiting satellite *had* come to grief over Siberia, could it, like the Vikings, have been attempting to actually set a life-seeking "lander" unit onto the surface of the Earth? Did it deliberately choose this barren, uninhabited area so that painstaking analysis of soil organisms and other ground-level research could go on undisturbed? In the Martianlike seclusion of the Southern Swamp it might have remained undetected for years, feeding masses of vital information to the computer "brain" of its orbiter, which, in turn, would have transmitted this to the planet of the alien star system from which it originated.

Though there could be many similarities in the self-sustaining functions of the Vikings and those of an interstellar satellite, the differences in the journeys they made would be vast. The first Viking took more than ten months to make the 200-million-mile trip to Mars before landing on the "Plain of Gold" on the sunny morning of July 20, 1976. A matter–antimatter–driven craft could cover the distance in about twenty minutes.

If therefore (and this is not inconceivable even within our own technological trends) a sophisticated matter–antimatter version of the Viking spacecraft was orbiting Earth in June 1908, let us now imagine the last moments of its fascinating mission. . . .

15

The Brain

The Brain moves in silence. A multitude of delicate sensors feeds a perpetual flow of information to the millions of microcomputer cells that form its membrane. Twenty-five thousand miles below, a pale-blue bauble is suspended among countless fairy lights in the endless blackness of space. It is the planet Earth.

The Brain's orders are to find life. They have been programmed into it by a supremely advanced intelligence somewhere out in the galaxy. It has come a long way. The incredible journey has taken many years. But its remarkable memory banks have been packed with masses of precise navigation data, geared to the changing positions of numerous star systems. Its artificial mind, so extensively programmed it can now think and act for itself, has been carefully shielded from the radiation and debris of outer space.

Six months ago it automatically swung into an orbit around the sun. Its sensors then systematically "interrogated" the solar planets for signs of those vital elements that could sustain life. They had soon selected Earth. Here was water and oxygen in abundance. The Brain had, therefore, made another maneuver and moved in for a closer probe, finally settling itself in a new orbit around this promising world.

The phenomenal energy force that has brought it here has been created by the awesome blending of matter and antimatter. Though almost all of this lethal fuel has been expended on its mammoth trek across space, enough of it remains safely isolated in powerful magnetic-force fields for any movements it might wish to make in our solar system.

Already the Brain has transmitted a detailed report on Earth to its own planet, though it will take many years to get there. It has analyzed the rich composition of the atmosphere. But there is no evidence of an advanced technology. If intelligent beings exist they are a comparatively backward race. There is no evidence of powerful radio signals, so there is little point in attempting long-distance contact. The amount of energy under their control is minimal. There are no indications of any exploitation of nuclear power.

The Brain considers all the relevant data. It has little to fear in making an even closer survey of the planet. It decides to dispatch a research lander onto Earth's surface, so it prepares to move into a tighter orbit. The calculations must be infallible, every movement precise. Already matching Earth's orbital speed, the Brain now orders that little extra boost necessary to overtake the planet and to nudge itself into its new "parking area."

Below sweeps the wrinkled face of Earth's ragged land forms, puckered and rutted and flaking into the vast blue oceans like multicolored peeling coats of paint that have been blistered and ravaged by its turbulent adolescence. It passes over the icing sugar of the North Pole that merges into the nooks and crannies of the frozen Canadian wastes. Now it is a soap-bubble world as all that the Brain "sees" is the enormous watery expanse of the Pacific.

One area dominates the attention of the geological- and biological-analysis cells in the Brain. It is the middle of Earth's greatest land mass, Eurasia. Sprawling in all di-

rections is the whole range of the planet's surface structure. Here is where intelligent life might flourish. This is
the region on which the investigation must be concentrated. It moves in closer. Too close. It is entering the
dense buffeting atmosphere of Earth.

The first indication of a malfunction in this masterpiece
of astroengineering is a "red alert" to the Brain, originating from the auxiliary fuel areas alongside the combustion
chamber of the craft. The magnetic-force field is breaking
up. An instantaneous series of system checks are fed
through the computer cells. Rapid countermeasures are
put into operation. They have no effect. The force field
continues to collapse.

The Brain battles with every permutation in its vast
program to control its flight path. But now it is streaking
deeper into the quagmire of the clawing atmosphere. Its
outer shell ignites like a fireball a thousand times its real
size. At any moment the deadly antimatter fuel will make
suicidal contact with the ordinary matter of the surrounding fuel chamber. The power supplies controlling the
other force fields die.

The moment of annihilation occurs at seventeen minutes and eleven seconds after seven in the morning five
miles above the dense taiga of central Siberia. With an
almighty flash of gamma rays an explosion vaporizes the
spacecraft in one dreadful microsecond of fury. Destruction is absolute. No vestige of it remains. "Mission Earth"
is over.

Nearly four hundred miles to the south, a train on the
Trans-Siberian Railway that snakes across the desolate
land below grinds to a halt. The passengers scamper onto
the track. For a few moments they listen to dull thuds
and distant rumblings. Then nothing more. The air is still
again. . . .

If in fact this fascinating scenario did occur in 1908,
it would have been some years before the Brain's last
dramatic message reached the intelligence that had dispatched it. Even if it had come from a planet of the

nearest star, Alpha Centauri, the radio "news flash" of its destruction would have taken more than four years to cover the journey across space. But, however long it took, the probe would have already transmitted enough pictures and data for an advanced intelligence to know that Earth was an ideal world to sustain life—that it was a place well worth further investigation. It is almost certain, therefore, that it would decide to send a second probe to carry on the interrogation started by its luckless predecessor.

But this time there would be no mistakes. The frantic stream of information on its failure that the Brain would have pumped out in those final dying seconds as it plunged Earthward would enable the alien scientists to produce a more sophisticated probe that would be programmed to cope with any similar emergencies.

So has a replacement probe already been sent here? Could it in fact be orbiting Earth at this very moment? It is at this point that the whole situation becomes *really* intriguing. Because in October 1973 the Russian astrophysicist Dr. Nikolai Kardashev and two other scientists at the Gorky Radio Institute, Academician Samuil Kaplan and Professor Vsevlod Troitsky, claimed they had actually picked up sequences of radio signals from space that *could* have originated from an alien intelligence. *And the signals appeared to be coming from somewhere within our own solar system,* which could only mean that they were being transmitted by an extraterrestrial probe.

The signals, claimed the Russian news agency Tass, occurred in "trains of pulses after definite lapses of time" and were "repeated over a wide range of wavelengths." Kardashev was later quoted in the American press as saying, "The character of the signals, their consistent pattern and their regular transmissions leave us in no doubt that they are of artificial origin—that is, they are not natural signals." Troitsky was reported as saying they were "definitely call signs from an extraterrestrial civilization."

These signals are said to have been detected simultaneously from widely separated areas of Russia, which completely eliminates local sources. The wavelengths on which they were transmitted are those that scientists would expect aliens to use if they were attempting to make contact. They were those that would not be unduly distorted by ionospheric disturbances. The sequence in which they were transmitted—specific trains of pulses each lasting a few minutes—is precisely the kind an orbiting probe would be likely to adopt.

No further confirmation or denial of these startling announcements by the Russians has been received by Western scientists. In July 1976 the Soviet press agency Novosti had still been sent nothing further on the signals despite its requests for information. We do not know if they were successfully decoded or whether others have been received since. We do know, however, that the search for extraterrestrial signals at Gorky and other places in the USSR has intensified since.

So, if these signals *were* alien—if they were not emanating from a runaway man-made satellite or some unidentified natural source—could they be coming from a probe sent to replace the one that exploded over Tunguska? For this time a computerized satellite would have no doubt that the world is inhabited by intelligent beings. There is now unquestionable evidence of this from the intense output of powerful radio signals transmitted into space from innumerable sources every moment of the day.

So the next question must be: Where could these probes have come from? There are 150 billion stars in our galaxy alone. How can we possibly begin to pick out the right one? Oddly enough, the task is not all that formidable. We can reduce the candidates to a very small number. From 1908, when we shall assume the original probe was destroyed over Siberia, to 1973, when Kardashev announced he had picked up artificial signals, it is sixty-five years. *Theoretically* this means that the farthest possible distance the planet could be from Earth is thirty-two and a half years. This allows thirty-two and a half years for the distress signal to get back from the first probe, and a further thirty-two and a half years for a second probe to reach Earth, traveling at the speed of light.

But, in *practice,* the maximum distance is considerably less. It is unlikely that a spaceship would travel at the speed of light. It would be extremely risky and uneconomical. The highest velocity an advanced race would be likely to consider for such a comparatively short interstellar journey is about half the speed of light. This would drastically cut down the amount of fuel (and consequently the space vehicle's original size) and also reduce the damaging effects of interstellar matter that might put vital and sensitive life-seeking instruments out of action, rendering the entire mission useless.

Travel at this more logical velocity narrows the choice of possible star systems to a twenty-light-year radius of Earth. This restricts us to eighty-one visible stars (see Appendix VII). But we can reduce this figure even further. For various reasons it is highly unlikely that the majority of these eighty-one stars are suitable to encourage the evolution of any kind of life form that we can conceive. However, within this twenty-light-year limit are *nine* star systems that certainly could be. And a number of these stars are known to have planets orbiting them. Some, too, are almost exactly like our own sun. This means that any form of intelligent life might be very similar to our own. Their systems could be based on carbon, water and oxygen. This does not necessarily mean they would *look* like us. As with automobiles, you can produce a vast range of conflicting designs, but their basic functions are the same. Communication and some acceptable level of compatibility would, therefore, be feasible. So let us now do a run-down on those nine star systems that would appear to have life-sustaining characteristics.

The nearest is the binary (double) system of Alpha Centaurus, 4.3 light-years away. It consists of two stars, Alpha Centauri A and B. Centauri A is virtually a twin of our own sun. It is in the same spectral class and 1.3

times as luminous. Its companion star, Centauri B, is about as close as the solar planet Saturn is to Earth. A Pioneer-type space vehicle could shuffle between the stars quite comfortably.

We can consider Alpha Centauri A for an example of how a man-made probe mission from Earth might work out. If the spaceship transporting it there accelerated at one Earth gravity (1g.) for one year, it would reach a speed of more than ninety percent that of light in a distance of half a light-year. It would then coast at this speed for another three years. *Deceleration* would then take a further one year, so that it would arrive in the region of Alpha Centauri A in about five years. A radio signal traveling back to Earth at the speed of light announcing its arrival would take 4.3 years, making a minimum total mission time of a little over nine years.

If the probe moved through space at *half* the speed of light, it would reach Alpha Centauri A in eleven years. Add the 4.3 years for its arrival message to get back to Earth, and we get a total mission duration of roughly fifteen years. This compares with a twenty-year mission time at thirty percent the speed of light and forty-eight years at ten percent.

An obvious question, if we consider Alpha Centaurus as the star system sending an Earth-bound probe, is why they would bother to do so when they are so close they could have been picking up our radio signals for nearly half a century. A possible answer is that they might wish to know far more about us through secret orbital surveys before rushing blindly ahead with direct communication. It is a basic instinct of all known life forms to approach any unfamiliar species with caution, learning as much about the other as they can before announcing their presence. Similarly, an alien race might want to know how advanced we were and whether we were the type of society that could be a threat to them in any way.

The next nearest star system worth considering is Epsilon Eridani, 10.69 light-years from our sun. This is one of the two stars on which Dr. Frank Drake, now director of the National Astronomy and Ionospheric Center at Arecibo, Puerto Rico, concentrated his search for extraterrestrial signals during "Project Ozma" in the early 1960s. Epsilon Eridani also appears to have planets. We can't *see* them, but astronomers can detect their presence either as they pass in front of the parent star, affecting its light output, or by the slight "wobbly" movement of the star caused by the gravitational pull of large planets orbiting it. Epsilon Eridani is believed to have at least one giant planet, about eight times the mass of Jupiter.

A little over eleven light-years from Earth is Epsilon Indi. Here life is just about credible, though the star is only thirteen percent as brilliant as our sun. About the same distance is Tau Ceti—the second star chosen for examination during Project Ozma. Tau Ceti is a very good candidate indeed for intelligent life: it is half as bright as our sun, and any planet orbiting it at about the distance of Venus (65 million miles) would produce ideal Earthlike conditions. A satellite probe traveling to Tau Ceti from Earth would take twenty-six years cruising at half the speed of light. Allowing almost another twelve years for an arrival signal to get back would make a total mission time of thirty-eight years. This would increase to fifty-five years moving at 30 percent the speed of light and 133 years at ten percent.

At almost seventeen light-years' distance is the star system called 70 Ophiuchi, suspected of having planets, and, at eighteen light-years, 36 Ophiuchi—twin stars with light outputs one quarter that of our sun and appearing to be quite capable of supporting Earthlike planets. In the same range are two other likely stars, Sigma Draconis and Delta Pavonis—the latter, like Alpha Centauri A, having a light output almost exactly that of the sun.

We now come to the final, and most exciting, star in the twenty-light-year radius for possible sources of probes since 1908. It is called Eta Cassiopeiae A and is 19.2 light-years away. In every respect it is virtually an identical twin of our sun. Its mass is similar, its light output is almost the same, and its companion star is known to have planets moving around it. This being the case, it is reasonably certain that Cassiopeiae A has planets, too, which would have been created from the same residual material when both these stars were first formed. It is so sunlike that, as explained earlier, a form of intelligent life supported by it could well be similar in many respects to our own. It could, therefore, well wish to investigate us. Another striking factor is that its lifetime is comparable to that of our sun, and it is very slightly brighter. This would produce a higher degree of radiation, which might result in a greater mutation rate than that which occurs on our own Earth. This being so, one would expect any intelligent beings there to have developed at a slightly faster rate. They would therefore be technically more advanced than mankind, and would be expected to be far better equipped in the field of interstellar travel and communication.

In every aspect Cassiopeiae A must be considered the most likely source of advanced intelligent life within twenty light-years of Earth. And the remarkable thing is that this star fits precisely into our time scale of sixty-five years (the period from the Tunguska explosion in 1908 to the first report of those Russian signals in 1973) if we are to consider a logical mission project.

This is how such a mission could have progressed:

June 1908: Probe explodes over Siberia and transmits final radio message back to Eta Cassiopeiae, 19.2 light-years away.

Mid-September 1927: Message received. It is apparent from the probe's earlier data that Earth is suitable for life. A second probe is justified. This takes between one and two years to plan, construct, test and launch.

1929: A spaceship carrying probe number two blasts off from Eta Cassiopeiae. It is designed to travel at half the speed of light—the most logical choice of velocity to cover the journey in a reasonable time and to give an acceptably efficient consumption of propulsion energy. To travel those 19.2 light-years—allowing five years for acceleration at launch and deceleration into the solar system —takes a total of 43.4 years.

1972: The probe settles in an orbit around Earth and begins transmitting signals. These are picked up by Kardashev at Gorky, and the first news is released by Tass in October 1973.

Conclusion

June 30, 1908, was a lifetime ago. Bustling new cities and sprawling industrial networks now grope across the once forlorn wilderness of Siberia, frisking its copious pockets of mineral wealth. Flourishing young communities are taking root where only the taiga grew. And armies of fresh young scientists regularly pound along the concrete routes that long since covered Kulik's pioneering tracks—all eager to pay homage at the overgrown graveyard of the monster that nature never fails to garnish with her continually replenished wreaths of botanical abundance. Like surgeons they prod and cut away the earth and amputate the trees, then scurry off back to space-age laboratories equipped with every conceivable technological refinement, still hoping that somewhere from these endless postmortems the true secret of the Southern Swamp will suddenly emerge.

Yet, despite the most searching techniques, despite a mighty reservoir of evidence and specimens that has accrued from half a century of exploration, analysis and debate, *some* gaping loophole, some flaw in an otherwise airtight solution, inevitably shows up. Still the mystery remains officially listed "unclassified." Nothing in

the continually widening field of known natural phenomena, it seems, can fully explain what really happened.

With each passing decade, the chances of solving the
riddle diminish. Even from the first soil samples taken
half a century ago, it is extremely difficult to identify any
telltale explosion materials from those that accumulated
naturally in the earth. In any case the vast proportion
of any chemical evidence would have vanished in
the explosion and been lifted and dispersed into the
stratosphere as dust, producing those bright skies over
many parts of the world in July 1908. What was left in
the ground offers few clues, either. The intense heat of
the blast melted the separate grains of sand and clay to
form those small balls or spherules of glass. It is virtually
impossible now for scientists to classify these.

As men continue to scoop up the soil of Tunguska,
NASA's mechanical Viking probe is doing precisely the
same thing on the boulder-strewn surface of Mars. It is
ironical that, though these computerized invaders millions
of miles from home are already sending us back the
secrets of a planet we have never even set foot on, our
most brilliant scientists still cannot unravel a mystery
right here on our own Earth. But perhaps, indirectly,
Viking is also offering the clue to Tunguska. Perhaps it
was also such a self-disciplined satellite that contained the
appalling power to devastate the taiga and all evidence
of itself in the awesome process.

But, whatever caused the Tunguska Event, there is the
inevitable question: Where and when will it happen
again? Certainly it must make us all aware of just how
vulnerable this tiny world of ours really is in the turbulent
universe. For how can we take even the most familiar
forces of nature for granted, let alone the many unknown
ones that lurk out there beyond the flimsy barrier of our
atmosphere? That seemingly empty space about us can
be ruthlessly unpredictable. We can never tell when some
friendly little glow in the night sky might once again
suddenly become the most destructive missile that man
has ever encountered.

On October 20, 1976, one such missile nearly collided with Earth. It was one of the nearest misses ever recorded. The object was an asteroid. Had it struck a populated area, the results would most certainly have been catastrophic. The asteroid, hundreds of yards across, passed 750,000 miles from our planet. That sounds a long way off. In astronomical terms it is frighteningly close. Astronomers coded it "1976 UA" and it was photographed at Mount Palomar as it sped by in its orbit round the sun. It will be centuries before this missile comes as close again. But remember, there are countless others where it came from.

Even our own sun, the very furnace of life, will one day reduce everything it has created to ashes. It will die. And in its tortuous death throes it will transform into an enormous "Red Giant" 250 times its normal diameter, slowly incinerating its entire family of planets. But that may not happen for billions of years. Perhaps the most immediate threat to life on this Earth is the unpredictability of man himself. For though no one knows what the thing that ravaged Tunguska really was, *everyone* knows of a brute that could cause the same kind of insane devastation. For man has already created his own Siberian monsters—enough of them to reduce the entire civilized world to one gigantic Cauldron of Hell. Caged in the nuclear reservations of the major powers, they are breeding at an alarming rate.

Perhaps the Tunguska Event is a timely message to us all.

APPENDIX I

Analysis of Tunguska Peat Samples
(See Chapter 5)

In 1971 the leading Soviet scientific journal *Doklady Akademii Nauk USSR* published the results of tests on peat at Tunguska by Dr. Yuri Dolgov and a team from the Meteorite and Cosmic Dust Commission of the Academy of Sciences. Dolgov dug samples to depths of eleven to fourteen inches. They included peat dating from 1908 containing residual material from the explosion. When this was isolated from the normal material by washing and burning the peat, it was found to consist of tiny glasslike spherules of various sizes. The largest were 1.4 millimeters (about one twentieth of an inch) in diameter. The smallest were 10 to 180 micromillimeters (about 4 to 80 micro inches). The largest were mostly transparent or colorless; the smaller ones were a variety of colors including blue and green. Some were black.

Their composition was unlike that of surrounding rocks and stones of the Podkamennaya Tunguska region. Neither did they have a makeup resembling moon rock samples. The closest comparisons the Russians could make for these spherules were to tektites—flat-shaped disks found in parts of Australia, Southeast Asia, Texas, the Ivory Coast and Czechoslovakia. These are mostly of a silica-rich glass (about seventy-five percent silica) with a composition close to granite. The true origin of tektites is unknown, but some scientists believe they may be the products of impacts by comets or giant meteorites on the Earth's surface. However, the resemblance was not entirely the same and the Tunguska material was richer in sodium and potassium. Concentrations of aluminum, silicon, calcium, magnesium and iron were present in both.

The Russians also analyzed *soil* at the Tunguska site which would contain traces of normal meteoritic and cosmic dust. These were not the same as the material found in the samples preserved in the peat. The soil had been contaminated by industrial products, whereas the peat samples remained uncontaminated, having been sealed between the original moss and that growing over it since 1908.

There is no mention in Dolgov's report of germanium—a material claimed to have been found in earlier analyses, and one that has been associated with the possible remains of a spaceship's semiconductors and transistors. In any case germanium can be found in ordinary meteorites. In the late 1960s, Drs. A. V. Jain and M. E. Lipschutz of Purdue University found evidence of germanium in 117 samples of meteorites from different areas. The amounts were within a narrow margin of about twenty parts per million.

Jain and Lipschutz believe most of their samples came from the Apollo family of asteroids, whose closest approach to the sun lies between Earth and Venus. Hermes and Adonis are two such "miniplanets." For each of these larger objects there are thousands more smaller ones, some merely a few feet across. Asteroids from this family are believed to have collided with Earth in the past and will certainly do so in the future. The result of a substantial one falling on a populous city would be catastrophic.

APPENDIX II

"Natural Nuclear Bombs"
(See Chapter 5)

Given the right ingredients and conditions, Nature herself can produce "H-bombs." It happens on an enormous scale in stars like our own sun, which throws off four million tons as light every second. A nuclear-type explosion might take place in the head of a comet by what could be termed "superfuels." These are perhaps a hundred times more powerful than ordinary gasoline or coal. They are not in themselves "nuclear" and do not directly produce nuclear explosions. But under extreme buildup of heat in a comet head entering Earth's atmosphere they may do so. This would be caused by the compression of the air building up in front of the comet, which acts as a brake on the fast-moving body. The air becomes white hot and superdense in the process. Anything trapped inside the comet cannot get out except by bursting the outer shell.

In one theory, comets are believed to be composed of solid frozen gases locked up in a strong stony outer casing. The Tunguska object might have been such a comet, its core consisting of frozen hydrogen ammonia mixed with liquid helium, in what is called a "triplet" state. Helium in this condition stores energy which is about ten to a hundred times more powerful than ordinary explosives. In addition it changes state and attempts to form ordinary gaseous helium at very high temperature (40,000 degrees C. at ordinary atmospheric pressure). If the pressure is raised to 125 atmospheres (1,800 pounds per square inch), the temperature approaches that required to detonate heavy hydrogen (about 5 million degrees C.). If the pressure rises to double this (3,600 pounds per square inch), the temperature produced by the triplet helium suddenly changing state may be enough to detonate ordinary hydrogen. The thick stony shell could possibly contain this pressure before it burst. Meanwhile the white-hot atmosphere would burn away the outer layer of the shell at the front where the shock waves were fiercest.

The sequence of events can, therefore, be summarized as follows: The comet enters the atmosphere, and the heat generated starts to "cook" the inside of the core and also to wear away the shell, which disappears as a cloud of dust trailing behind. Eventually the tightly packed core reaches a critical temperature where the trapped gases try to change state. They cannot, because the thick shell prevents this. Up goes the temperature caused by the explosively changing state of the internally trapped gases. If the outer shell can hold together up to the correct critical pressure, then the temperature is forced to a point where any trapped hydrogen detonates or fuses to form helium, as in a normal H-bomb. The amount of hydrogen needed to produce a thirty-megaton explosion is not large (about 2,000 pounds), for a proportion of the energy will come from the other parts of the pent-up gases being released.

The detonation, of course, would destroy most of the stony shell, though some of it might be melted into silicate spherules. Those found at Tunguska are said to have some resemblance to tektites (see Appendix I).

These are a rocky glass material similar to Earth granite. Some forms of granite might be tough enough to form a shell able to contain up to 3,600-pounds-per-square-inch pressure.

APPENDIX III

Black Holes and Ball Lightning
(See Chapters 6 and 7)

Dr. Steven Hawking (*Nature,* Vol. 248, 1974) has produced a general formula for small black holes showing that below a critical limit of half the Earth's mass a hole cannot sustain itself and will exist for only a geologically short time (a million years at the most). Smaller holes have shorter lives. This is a serious flaw in associating the phenomena with the Tunguska Event. For such an encounter with Earth, the black hole must be expected to have been in existence for a reasonably long period. If so, it would be at or near Hawking's limit. In this case it would demolish the Earth on impact.

When considering the possibility of "micro black holes" being created by the action of lightning, it is possible to estimate the size such a hole would have to be to correspond with the lifetime of only a few minutes or seconds.

Professor John Taylor of King's College, London, in his book *Black Holes,* calculates that the energy obtainable from a single black hole has a theoretical upper limit of thirty percent of the total rest mass. In other words, $E = \dfrac{mc^2}{3}$. This is at least one hundred times more efficient than fusion energy. But let us assume that a micro-black-hole lightning ball extracts only five to ten percent of the energy of the mass. We can then calculate the mass of a black hole sufficiently explosive to demolish an average-sized room. This would roughly correspond to an explosive charge of one to ten kilograms (2.5 to 25 pounds) of TNT. So how much matter is required to produce this force when used at an efficiency of ten percent?

Scientists use a unit called the erg when relating mass and energy. In the case of the Tunguska explosion we are dealing with 10^{24} ergs of energy, producing the equivalent of thirty megatons of TNT. If we scale this down, we get the reduced comparison of one kilogram of TNT explosive being equal to 5×10^{13} ergs. Ten kilograms of TNT will equal 5×10^{14} ergs. We can now work out the equivalent mass of a micro black hole.

The rest mass energy of 1 kilogram of any matter is 10^{24} ergs. To get 5×10^{13} ergs, we require only 5×10^{-11} kilograms if we have 100 percent efficiency. But we are assuming the micro black hole is only five to ten percent efficient in obtaining energy. The mass is therefore larger (about 10^{-9} kilograms).

A tiny fraction of matter compressed into a micro black hole produces the explosion of ball lightning. The mass involved is one billionth of a kilogram, or roughly half a billionth of a pound of mass. Since this mass would be tremendously compressed, the diameter of the sphere into which it is contained works out at 4×10^{-27} meters, which is smaller than the diameter of a hydrogen atom.

It can be estimated that a one-foot-wide lightning ball or Kugelblitz has a total energy force of 10 million joules of which it dissipates some twenty percent, or 2 million joules. A joule is equal to 10^7 ergs. Two million joules equals 2×10^{13} ergs. Lawton now estimates the size of a Kugelblitz capable of causing the Tunguska devastation. Assuming that the energy is proportional to the volume of the explosive (whatever that was), then the volume of the Siberian object would be larger by the ratio of 10^{24}

divided by 2 $\times$ 10^{13}—or 5 $\times$ 10^{10} larger. The volume of a sphere is proportional to the cube of its diameter, so the diameter of the sphere would be the cube root of 5 $\times$ 10^{10} larger than the one-foot Kugelblitz. This works out at 3,700 feet, or a little over one kilometer.

It has been suggested that ball lightning is composed of dust bound tightly together by an electrical charge. Professor Lyttleton of the Mathematics Department at Cambridge University firmly believes that comets too are large dust balls. A likely force in space that might bind the dust into a ball could be electrostatic charges. These could be generated by the solar wind particles "rubbing" the outer layers of the dust and debris. The wind particles are small and, in the outer parts of the solar system, are not intense enough to disrupt the comet as may happen when it sweeps toward the sun. The "rubbing" process produces effects similar to those from polishing a piece of amber, causing it to attract dust or bits of paper.

The electrical charge in a comet cloud can be enormous —literally billions of volts. There is no way for the charge to leak away in the insulation of space. If a one-kilometer electrostatically charged Kugelblitz entered the Earth's atmosphere, it is just possible it would disintegrate in a giant electrical flash. This could produce the scorching radiant energy, the searing shock waves and the negative pressure effect (NPE) referred to in Chapter 9.

APPENDIX IV

Notes on Antimatter
(See Chapter 7)

The simplest possible atom, that of hydrogen, consists of a central core or nucleus with only one proton. Around this central core spins a single electron, much like a lone star with only one planet orbiting it. Traditionally, the electron is associated with a negative electrical charge, which means that the proton must have an equal and opposite positive charge if the atom itself has zero charge. In other words, the opposing charges of the proton and the electron cancel each other out.

In 1933 it was suggested that there should be "anti-particles" which were mirror images of the conventional ones, the central proton being *negative* and the electron positive, thus still preserving an overall zero charge as in the case with ordinary material. This speculation was confirmed when the "positive electron," or "positron," was discovered in 1934. Since then a number of other anti-particles have been recognized—such as antiprotons (sometimes called negative protons), antineutrons, antineutrinos, etc.—as higher and higher energies are used to smash up atoms and their flying debris is examined.

The following list is the atomic composition of a few well-known materials referred to in the main body of this book, together with their antimatter counterparts. In each case the number of protons, neutrons and electrons contained in the normal matter are equal to the negative protons, antineutrons and positrons of the antimatter. For example: carbon has 6 protons, 6 neutrons and 6 electrons; anticarbon, therefore, has 6 negative protons, 6 antineutrons and 6 positrons. Other combinations are:

Oxygen and antioxygen (8, 6, 8)

Silicon and antisilicon (16, 16, 16)

Uranium 235 and anti-U235 (92, 143, 92)

Plutonium 242 and antiplutonium 242 (94, 148, 94)

Hydrogen has only one proton and one electron, giving antihydrogen one negative proton and one positron.

If we attempt to mix any of the above pairs—or indeed any other material with its *exact* mirror image—it would invariably result in the same enormous release of gamma-ray energy. But what would happen if we mixed *different* types of matter–antimatter elements—for example, silicon with anticarbon? (Silicon has 16 protons, 16 neutrons and 16 electrons, whereas anticarbon has only 6 negative protons, 6 antineutrons and six positrons.) After the initial gamma flash we would theoretically be left with a material that had 10 protons, 10 neutrons and 10 electrons. No such material exists. Instead the residue would sort itself out so that some of the protons would swiftly change into neutrons, the positrons given off decaying in gamma radiation by meeting stray electrons.

A likely material remaining would be ordinary carbon (6 protons, 6 neutrons, 6 electrons), leaving 4 protons, 4 neutrons and 4 electrons over, which conveniently splits into two atoms of deuterium (heavy hydrogen)—i.e., two sets of 2 electrons, 2 neutrons and 2 protons. We would, in fact, have lost the antimatter in the energy flash and produced a residue of ordinary matter which may or may not be similar to the original. In the above example we would be left with ordinary carbon and ordinary heavy hydrogen, more correctly termed deuterium.

APPENDIX V

Notes on Nuclear Reaction
(See Chapter 7)

The nuclear physicist looks for processes which involve a loss of mass between elements at the start of the process and those left at the finish. The classical case is that of U235, which splits up naturally and of its own accord. In doing so, the massive uranium atom splits into two unequal parts. One part is the metal barium of mass 139; the other is the metal molybdenum of mass 95. Therefore the mass left over is $\frac{234}{235}$ or 0.996 of the original. This means that the tremendous explosion release in an atom bomb is caused by the loss of only 0.4 percent of the mass.

When dealing with nuclear reactors, the reaction rate is slowed down and the explosive chain reaction kept to a continuous but controlled heat. In early reactors the temperature was nearly 400 degrees C. (black heat), but in later models the working temperatures are much higher. The total amounts of working material in some of these reactors is large, but the actual matter destroyed is still extremely small (about 0.05 to 0.2 percent of the original weight). In practical terms, the size of a block of uranium used to drive a nuclear submarine round the world is roughly equal to half a house-brick. We would be left with a mixture of barium and molybdenum metals whose size would be slightly smaller than the original block of uranium. But the amount of actual matter that had vanished would be roughly equivalent to a couple of sugar lumps.

Definition of "Half-life"
(*See Appendix VI*)

"Half-life" is a term used by the nuclear physicist to determine how long a radioactive material may exist. Most of the ninety-two natural elements in the Periodic Table are stable, but a few of them are naturally radioactive. This means they decay by giving off particles or rays and transform into different elements. For example, uranium 235, the metal normally associated with nuclear bombs, will, if left to itself, gradually decompose to form lead. Therefore if, at the present time, we have one pound of U235, then at some time in the far distant future we should have slightly less than one pound of lead. The balance of weight lost will have disappeared as energy. But how long will we have to wait?

U235 has a half-life of 71 million years. This means that after that period we would have lost half our original weight of uranium (i.e., half a pound). In a further 71 million years we would have lost another quarter of a pound, then a further one eighth of a pound, and so on. In 213 million years, seven eighths of the U235 would have turned to lead, and after another 213 million years 63/64ths of it would be lead.

It should be noted that U235 (unless stacked in large quantities) is not *violently* radioactive material. However, radium, the element discovered by Pierre and Marie Curie, is an extremely radioactive element. It can assume a variety of forms, the most active of which is radium 224, which quickly decays to thorium in just over three and a half days. The variety discovered by the Curies and used for the treatment of cancer is radium 226, which has a half-life of 1,600 years. Even so, this is short, and accounts for the struggle the Curies had in obtaining this rare and valuable element. From a ton of crude pitchblende ore they obtained a few micrograms of radium— enough to discolor the bottom of the dish holding it. But the radioactive glow from the stain lit up their laboratory and badly burned the hands of Marie Curie and other pioneer radiologists, making them the first radiation "victims."

Some materials have half-lives in millionths or billionths of a second, others in billions of years. However, they are still radioactive and must not be confused with the entirely stable elements that will last indefinitely.

APPENDIX VI

Tests for Antimatter at Tunguska
(See Chapters 7 and 14)

Dr. Hall Crannell (*Nature,* Vol. 248, pp. 396, 397) points out that a matter–antimatter explosion should produce aluminum 26 (symbol ^{26}A1) by interaction with silicon. The most common form of silicon is silicon 28 (symbol ^{28}Si). In such an explosion, a proton and an antiproton annihilate and produce an average of four charged pions (pi-mesons) and two neutral pions. If the silicon 28 absorbs *one* charged pion, is will become ordinary aluminum 27, which is not radioactive and not detectable. However, if it absorbed *two* pions (and there would be plenty around in such an explosion), it would then become aluminum 26, which *is* radioactive and, therefore, can be detected.

Dr. Crannell goes on to illustrate the type of counter-system that could be used and then shows that, with aluminum 26 having a radioactive half-life of 750,000 years (see Appendix V), and the number of aluminum-26 atoms in the Tunguska site center estimated as 2×10^9 per pound of rock, there should be an extremely good chance of finding such material should a matter–anti-matter explosion have occurred. Crannell estimates that in such an explosion the dosage concentration of ^{26}A1 should be fifteen percent higher than the normal background dosage received by the surrounding area in one million years. Since his countersystem can detect an increase as small as a few percent (say less than five), he feels confident that this test could confirm whether the explosion was due to antimatter.

His test has not yet been carried out, but we have been able to examine the results of the mineral composition of small silicate spherules obtained from the impact area. The composition is quite different from that of normal rock. Crannell considered his average rock as having 29 percent silicon and 8 percent aluminum, but our information shows that the Tunguska material has a silicon content of approximately 25 percent and an aluminum content of 2 percent, the balance largely being oxygen and traces of iron and manganese. With silicon/aluminum ratios much higher than Earth rocks, it is highly probable that the aluminum *will* have a high percentage of ^{26}Al should it be the residue of a matter-antimatter explosion.

The current (1976) expedition dispatched by the USSR Academy of Sciences will almost certainly collect rock and soil samples from the impact area. If they subject these to the Crannell test and do find high levels of ^{26}Al (in which case the antimatter theory is acceptable), then the most likely source of the material would be *artificial* rather than natural (see Chapter 14). Natural antimatter would have great difficulty in surviving contact with dense particles in the solar system. The sun too emits a continuous stream of solar-wind particles and is itself enveloped in a cloud of dust which reaches out beyond the asteroid belt and shows as the zodiacal light. This dust and gas are relatively harmless to ordinary objects (including our own spacecraft), but if an antimatter object entered the solar system it would be continuously attacked and gradually eroded in a tremendous outpour of energy. This would produce the bright blue "tube" seen by witnesses in Siberia.

If antirock did reach the ground and bury itself, there

would be the most frightful explosion, certainly capable of causing the damage recorded. However, a ground explosion would leave a crater several kilometers across, and seismographs in various parts of the world would have recorded appropriate *crustal* earthquake shock, not the surface waves which were reported.

Antimatter Used as Rocket Fuel
(See Chapter 14)

Antimatter material used as a rocket fuel would be reacted with ordinary matter to form pure radiant energy. If this was focused in some way not as yet developed by our technology, the most conceivably efficient interstellar propulsion system would result. With a matter–antimatter rocket capable of traveling at 90 percent the speed of light, every ton of payload would require 34 tons of fuel—i.e., an equal amount of matter and antimatter. A Saturn rocket has to carry 1,200 tons of conventional fuel for every ton of payload to attain a speed of only 25,000 miles an hour.

APPENDIX VII

Nearest Stellar Targets for Automated Probes,
and the Use of Antimatter-Energized Propulsion
(See Chapter 15)

The following are excerpts from *A National Space Program for Interstellar Exploration,* by Dr. Robert L. Forward, Hughes Research Laboratories, Malibu, California, submitted at the invitation of the Subcommittee on Space Science and Applications, Committee on Science and Technology, United States House of Representatives, July 1975.

Within twenty light years of the Sun there are 59 stellar systems containing 81 visible stars. These include 41 single stars, 15 binary and three triple systems.

First-generation probes will search for planets existing in a temperature zone roughly equivalent to the Venus-Jupiter zone in the Solar System. This includes not only planets in single star systems, but also small satellites of large Jovian-type planets orbiting within the correct temperature zone of the primary, and those planets in binary (and triple) systems whose total incident energy flux raises surface temperatures to the correct levels (Huang, 1960). Also included are those planets with highly eccentric orbits, which spend only part of their orbital period within the correct zone.

Since one star, our Sun, is positively known to possess a planetary system, it can only be assumed that stars similar to the Sun in spectral type, mass, radius and luminosity also have some possibility of possessing a planetary system. By using a star's similarity to the Sun as a selection criterion, we hopefully will be choosing target stellar systems not only for their potential planetary systems but also on their ability to support life.

The Sun is a GO-type star with an effective temperature of ∼5900 degrees K. Unfortunately there is only one other GO star (Alpha Centauri A) and a total of only four G-type stars within 20 light-years. The selection criterion can be extended to include all the cooler K-type stars. Below these lie the red dwarfs (M-type), generally regarded as being too small and cool to supply the required amount of energy for life to any planet, except one in an extremely close orbit. This has led many to eliminate red dwarf stars as prospective targets for interstellar probes. However, we know red dwarfs have planets (six of the eight stars known to have planets are red dwarfs) and until we are sure that a system has no planet within its life-supporting zone, we should not eliminate it from consideration as a potential target for an interstellar probe.

Since more than 25 percent of the stellar systems under consideration are multiple, the possibility of a planet occuring in a stable orbit and a temperature zone conducive to life in a multiple system must be considered. For a while it was thought that in most multiple systems the formation of several stars simultaneously prevented the formation of a planetary system. However, at least two binary systems, 61 Cygni, and Eta Cassiopeiae, are known to have planetary companions.

For missions to the nearby stars, we do not need vehicle velocities very close to that of light, so we do not need the exhaust velocity to be that of light. A more efficient method is to use a small amount of antimatter to energize a much larger amount of regular matter (Papailiou, 1975).

An optimized engineering study of an interstellar propulsion system using antimatter has yet to be done, but some approximate numbers can be estimated now. If we assume the development of a very light weight (10kg.) interstellar exploration probe design, and an antimatter propulsion system optimized for a probe coast velocity of one-third the speed of light, then the minimum in the mass ratio of each propulsion stage occurs at a mass ratio of 4 (Papailiou, 1975), and the antimatter mass required is 1/10th of the burnout mass, or 2 percent of the initial mass. The interstellar probe at launch would consist of a 51-kilogram second stage, 200 kilograms of propellant and 5 kilograms of antimatter. This would suffice to accelerate the second stage to one-third the velocity of light. The 51-kilogram second stage would consist of the 10-kilogram exploration probe payload, 40 kilograms of propellant and 1 kilogram of antimatter which would be used to decelerate the probe at the target star.

Besides the obvious engineering problems that remain to be solved in the design of an engine that can effectively convert the antimatter energy into directed thrust of the propellant, there is a real question of the obtainability and control of the antimatter.

The containment and control, once made, should not be too difficult since we have a number of ways of applying forces to the antimatter without touching it. Electric fields, magnetic fields, rf fields and laser beams are all used in present-day technology to levitate and control small amounts of regular matter that we do not want to contaminate. These would all be equally effective on antimatter.

The generation of appreciable amounts of antimatter is the primary engineering problem that should be addressed early in any investigation of the feasibility of using antimatter for interstellar propulsion. We already destroy hundreds of kilograms of regular matter each year in our electric generating plants and convert it to electricity. All we need to do is to develop efficient methods of turning that energy back into antimatter instead of regular matter.

The present methods for producing antimatter involve the use of large linear accelerators which can produce a proton beam of 10^{15} protons per second. When such a beam collides with a target, antiprotons are produced as part of the debris. The antiproton yield is independent of beam energy and is of the order of one percent. Factors such as limitations in the number of antiprotons which can be captured by an applied magnetic field and optimum size of the target in order to avoid collisions of antiprotons with target nuclei reduce further the production of antiprotons by two orders of magnitude. Hence, the rate of antiproton production is of the order of 10^{11} antiprotons per second. Based on this number, the estimated time to produce one kilogram of antiprotons from one machine is 10^8 years. This number has led some people to the conclusion that the use of antimatter for propulsion purposes is not feasible.

However, the presently used methods are not designed for antimatter production, but rather for studies in the physics of elementary particles. A study to investigate the possibility of increasing antimatter production rates to the level required for propulsion applications would be one of the critical technology areas that should be studied in the initial phases of an interstellar exploration program. The problems of antimatter storage, annihilation rate control and development of practical propulsion schemes are also those on which future studies should be focused.

BIBLIOGRAPHY

Altschuler, M. D., House, L. L., and Hilder, E., "Is Ball Lightning a Nuclear Phenomenon?," *Nature,* Vol. 228, p. 545 (1970).

Amazing World of Nature, The. The Reader's Digest Assn., 1969.

Ambartsumyan, V., *Soviet Sputniks,* Soviet Booklet No. 25.

Anderson, D. L., "The San Andreas Fault," *Scientific American,* November 1971.

Astapovich, I. S., "Air Waves Caused by the Fall of the Meteorite of 30 June 1908 in Central Siberia," *Quarterly Journal of the Royal Meteorological Society,* 1934, No. 2.

———, "More about the Night of 30 June 1908," *Mirovedeniye,* 1926.

———, in *Soviet Astron,* Vol. 10, p. 465 (1933).

Atkins, T., and Baxter, J., *The Fire Came By.* Doubleday, 1976.

Beasley, W. H., and Tinsley, B. A., "Tungus Event Was Not Caused by a Black Hole," *Nature,* Vol. 250, pp. 555, 556 (1974).

Bigg, E. K., "Influence of the Planet Mercury on Sunspots," *Astronomical Journal,* Vol. 72 (1967).

Bosanquet, F. C. T., *Pliny's Letters.* London: George Bell & Sons, 1907.

Boyarkina, A. P., and Bronsten, I. S., "Ob. energii vzryva Tungusskogo meteorita i uchete neodnorodnosti atmosfery," *Astronomicheskii vestnik,* Vol. 8, No. 3 (1975), pp. 172–77 (English abstract).

Brown, P. L., *Comets, Meteorites and Men.* Robert Hale, 1973.

Burns, J. O., Greenstein, G., and Verosub, K. L., "The Tungus Event as a Small Black Hole," *Monthly Notices, Royal Astronomical Society,* Vol. 175, pp. 355–57 (1976).

Charman, N., "Ball Lightning Photographed," *New Scientist,* February 26, 1976.

———, "The Enigma of Ball Lightning," *New Scientist,* December 14, 1972.

Comets. Ash & Grant, 1973.

Crammer, J. L., and Peierls, R. E., *Atomic Energy.* Penguin, 1950.

Crannell, H., "Experiments to Measure the Antimatter Content of the Tunguska Meteorite," *Nature,* Vol. 248, pp. 396–98.

Daily Telegraph, London, March 21, 1964, "Invitations from Outer Space."

Daily Telegraph Magazine, July 1976, "Newest Merchandise of Doom."

Dole, S., and Asimov, I., *Planets for Man.* Rand Corp., 1964.

Dolgov, Y. A., Vasil'ev, N. V., L'vov, Y. A., *et al.,* "Chemical Composition of Silicate Spherules in Peats of the Tunguska Meteorite Fall Region," *Doklady Akademii Nauk USSR,* Vol. 200, No. 1 (1971), pp. 201–4.

Feis, H., *The Atomic Bomb and the End of World War II.* Princeton University Press, 1966.

Fesenkov, V. G., "The Nature of the Tunguska Meteorite," *Meteoritika,* Vol. 20, pp. 27–31 (1961).

Fleming, S. J., and Aitken, M. J., "Radiation Dosage Associated with Ball Lightning," *Nature,* Vol. 252, pp. 220–21 (November 1974).

Florensky, K. P., "Preliminary Results from the 1961 Combined Tunguska Meteorite Expedition," *Meteoritika,* Vol. 23, pp. 3–29 (1963).

Galanopoulos, A. G., "Die ägyptischen Plagen und der Auszug Israels ans geologischer Sicht," *Das Altertum,* Vol. 10 (1964).
Gass, I. G., *Understanding the Earth.* Cambridge (Mass.): MIT Press, 1972.
Gregory, J. S., *Russian Land Soviet People.* George G. Harrap, 1968.
Griggs, R. F., *The Valley of Ten Thousand Smokes.* National Geographic Society, 1922.
Glasstone, S., *The Effects of Nuclear Weapons.* U.S. Atomic Energy Commission, 1962.
————, *Nuclear Reactor Engineering.* Macmillan, 1956.
Gribbin, J., and Plagemann, S., *The Jupiter Effect.* Macmillan, 1974.

Heilprin, A., *Mount Pelée and the Tragedy of Martinique.* Philadelphia: J. B. Lippincott, 1903.
Hibbert, C., *London: The Biography of a City.* Longmans, Green, 1969.
Hindley, K., "Comets in Perspective," *New Scientist,* April 17, 1976.
Hirschfeld, B., *A Cloud over Hiroshima.* Bailey Bros. & Swinfen, 1974.
Hughes, D. W., "Earth—an Interplanetary Dustbin," *New Scientist,* July 8, 1976.

Ince, M., "A Meteorite Crater in Britain," *Hermes,* July 1972.

Jackson, A. A., and Ryan, M., "Was the Tunguska Event Due to a Black Hole?," *Nature,* Vol. 245, pp. 88–89 (1973).
Jordan, E. C., *Electromagnetic Waves and Radiating Systems.* Prentice-Hall, 1970.
Jungk, R., *Children of the Ashes.* William Heinemann, 1961.

Kardashev, N. S., "Transmission of Information by Extraterrestrial Civilisations," *Astronomical Journal,* Vol. 41 (1964).

Kennan, G., *Siberia and the Exile System.* New York: Century Co., 1891.

———, *The Tragedy of Pelée.* New York: The Outlook Co., 1902.

King, J. W., "Solar Radiation Changes and the Weather," *Nature,* Vol. 245, p. 443 (1973).

Krinov, E. L., *Giant Meteorites.* Pergamon, 1966.

———, *Principles of Meteoritics.* Pergamon, 1960.

———, "The Siberian Meteorite Fall of February, 1947," *Sky and Telescope,* Vol. 15, No. 7 (1956).

———, "The Tunguska Meteorite," *Publ. Acad. Sciences USSR,* 1949, p. 196.

———, "Der tungusker Meteorite," *Chemie der Erde,* Vol. 20, No. 3 (1958).

Kulik, L. A., "Account of Meteorite Expedition," *Journal of the Russian Academy of Sciences,* 1922.

———, "The Lost Filimonovo Meteorite of 1908," *Works of the Lomonossoff Institute of the Russian Academy of Sciences,* Vol. 10, No. 1 (1921).

———, "On the History of the Bolide of 1908 June 30," *Journal of the Russian Academy of Sciences,* 1927.

———, "Preliminary Results of the Meteorite Expeditions Made in the Decade 1921–31," *Works of the Lomonossoff Institute of the Russian Academy of Sciences,* 1933, Part II, pp. 73–80.

Lawton, A. T., "The Interpretation of Signals from Space," *Spaceflight,* Vol. 15 (1973).

———, "Interstellar Communication—Antenna or Artifact," *Journal of the British Interplanetary Society,* April 1974.

Marinatos, S., "Massive Meteorite Blast in Siberia Really Nuclear Explosion," *National Enquirer,* May 13, 1975.

———, "The Volcanic Destruction of Minoan Crete," *Antiquity,* 1939.

Mattinson, H. R., "Project Daedalus: Astronomical Data on

Nearby Stellar Systems," *Journal of the British Interplanetary Society,* Vol. 29, No. 2 (1976).

Mavor, J. W., Jr., *Voyage to Atlantis.* Putnam, 1969.

McCall, G. J., *Meteorites and Their Origins.* David and Charles, 1973.

Melmoth, W., *The Letters of Pliny the Consul.* London: R. & J. Dodsley, 1763.

Millman, P. M., *Meteorite Research.* Dordrecht: D. Reidel, 1969.

National Enquirer, July 15, 1975, "Secret Evidence Shows UFOs Come from Other Worlds."

Nature, Vol. 247, p. 423 (1974), "Meteorites Which Bounce off the Earth."

New Scientist, April 15, 1976, "New Japanese Report Registers Grim Atom Bomb Toll."

————, April 29, 1976, "Centaurus-A Powered by a Massive Black Hole."

————, May 6, 1976, "World's Largest Stone Meteorite Falls on China."

Nice-Matin, October 11, 1972, "Le Mystère de Montauroux."

————, October 12, 1972, "Montauroux."

Nininger, H. H., *Our Stone-Pelted Planet.*

Obruchev, S. V., "Concerning the Place of the Fall of the Large Khatanga Meteorite of 1908," *Works of the Lomonossoff Institute of the Russian Academy of Sciences,* Vol. 10, No. 1 (1921).

Parry, A., *Russian Rockets and Missiles.* Macmillan, 1961.

Petrov, G. I., and Stulov, V. P., "Novaya gipoteza a Tungusskum meteorite," *Zemlya i Vselennava,* 1975, No. 4, pp. 74–75.

Polkanov, A. A., "Phenomena Accompanying the Fall of the Tunguska Meteorite," *Meteoritika,* 1946, No. 3, p. 69.

Pompeii, the Rediscovered City. Novara: Instituto Geografico de Agostini, 1969.

Project Daedalus, Interim Report of the British Interplanetary Society Starship Study.

Rawcliffe, R. D., "Meteor of August 10, 1972," *Nature,* Vol. 247, pp. 449, 450 (1974).

Ricketts, A., *Fundamentals of Nuclear Hardening of Electronic Equipment.* Wiley, 1972.

Rondière, P., *Siberia, Land of Promise.* Constable, 1962.

St. George, G., *Siberia: The New Frontier.*

Sobolev, V. S., ed., *The Problematic Meteorite.* USSR Academy of Sciences, 1975.

Sorlin, P., *The Soviet People and Their Society.* Pall Mall Press, 1968.

Spaceflight, July 1976, "NASA on Interstellar Flight."

Stadling, J., *Through Siberia.* Constable, 1901.

Stern, D. K., "First Contact with Non-Human Cultures," *Mercury,* journal of the Astronomical Society of the Pacific, September/October 1975, pp. 14–17.

Stewart, W., *Characters of Bygone London.* George G. Harrap, 1960.

Stoneley, J., and Lawton, A. T., *CETI, Communication with Extra-terrestrial Intelligence.* Warner Books, 1976; W. H. Allen (Starbooks), 1976.

———, *Is Anyone Out There?* Warner Books, 1974; W. H. Allen (Starbooks), 1975.

Sullivan, W., "A Hole in the Sky," *New York Times Magazine,* July 14, 1974.

———, *We Are Not Alone.* Hodder & Stoughton, 1965.

Sunday Times, London, April 4, 1976. "Mutation Mystery."

Suslov, I. M., "In Search of the Great Meteorite of 1908," *Mirovedeniye,* Vol. 16 (1927).

Symons, G. T. *The Eruption of Krakatoa and Subsequent Phenomena.* London: Krakatoa Committee of the Royal Society, 1888.

Taylor, J., *Black Holes.* Souvenir Press, 1973.

Times, The, London, June 29–July 10, 1908.

Tompkins, D. R., and Rodney, P. F., "Photographic Observation of Ball Lightning," *Soviet Physics JETP Letters,* Vol. 18, p. 114.

Troitsky, V. S., and Kaplan, S. A., reported work, Novosti bulletin, October 1973.

————, reported work, Tass, October 16, 1973.

Valéry, N., "The Shape of War to Come," *New Scientist,* June 17, 1976.

Vasil'ev, N. V., Ivanova, G. M., and L'vov, Y., "New Data on the Composition of the Tunguska Meteorite," *Prirodo,* 1973, No. 7, pp. 99–101.

Vasil'ev, N. V., L'vov, Y., *et al.,* "Silicate Spherules in Peats of the Tunguska Meteorite Fall Region," *Doklady Akademii Nauk USSR,* Vol. 199, No. 6 (1971), pp. 1400–1402.

Voznesensky, A. V., "The Fall of the Meteorite of June 30, 1908, on the Upper Part of the River Khatanga," *Mirovedeniye,* Vol. 14, No. 1 (1925).

Whipple, F. J. W., in *Quarterly Journal of the Royal Meteorological Society,* Vol. 16, p. 287 (1930).

Wick, G. L., and Isaacs, J. D., "Tunguska Event Revisited," *Nature,* Vol. 247, p. 139 (1974).

Wilcoxson, K. *Volcanoes.* Cassell, 1966.

Wood, K. D., "Sunspots and Planets," *Nature,* Vol. 240, No. 91 (1972).

Yavnel, A. A., "Meteoritic Matter from the Place Where the Tunguska Meteorite Fell," *Astronomical Journal,* Vol. 34, pp. 794–96.

Zigel, F., "Nuclear Explosion over the Taiga," *Znaniya-Sila,* Vol. 12 (1961).

Zolotov, V., *Doklady Soviet Phys.,* Vol. 12, p. 108 (1967).

Index